T0224411

Lecture Notes in Computer Science   8980

Commenced Publication in 1973
Founding and Former Series Editors:
Gerhard Goos, Juris Hartmanis, and Jan van Leeuwen

Editorial Board

David Hutchison
  Lancaster University, Lancaster, UK
Takeo Kanade
  Carnegie Mellon University, Pittsburgh, PA, USA
Josef Kittler
  University of Surrey, Guildford, UK
Jon M. Kleinberg
  Cornell University, Ithaca, NY, USA
Friedemann Mattern
  ETH Zurich, Zürich, Switzerland
John C. Mitchell
  Stanford University, Stanford, CA, USA
Moni Naor
  Weizmann Institute of Science, Rehovot, Israel
C. Pandu Rangan
  Indian Institute of Technology, Madras, India
Bernhard Steffen
  TU Dortmund University, Dortmund, Germany
Demetri Terzopoulos
  University of California, Los Angeles, CA, USA
Doug Tygar
  University of California, Berkeley, CA, USA
Gerhard Weikum
  Max Planck Institute for Informatics, Saarbrücken, Germany

More information about this series at http://www.springer.com/series/8637

Abdelkader Hameurlain · Josef Küng
Roland Wagner · Hendrik Decker
Lenka Lhotska · Sebastian Link (Eds.)

Transactions on Large-Scale Data- and Knowledge-Centered Systems XVIII

Special Issue on Database- and Expert-Systems Applications

Springer

Editors-in-Chief

Abdelkader Hameurlain
IRIT, Paul Sabatier University
Toulouse
France

Roland Wagner
FAW, University of Linz
Linz
Austria

Josef Küng
FAW, University of Linz
Linz
Austria

Guest Editors

Hendrik Decker
Instituto Tecnológico de Informática
Valencia
Spain

Sebastian Link
The University of Auckland
Auckland
New Zealand

Lenka Lhotska
Czech Technical University
Prague
Czech Republic

ISSN 0302-9743   ISSN 1611-3349 (electronic)
Lecture Notes in Computer Science
ISBN 978-3-662-46484-7   ISBN 978-3-662-46485-4 (eBook)
DOI 10.1007/978-3-662-46485-4

Library of Congress Control Number: 2015932684

Springer Heidelberg New York Dordrecht London

© Springer-Verlag Berlin Heidelberg 2015
This work is subject to copyright. All rights are reserved by the Publisher, whether the whole or part of the material is concerned, specifically the rights of translation, reprinting, reuse of illustrations, recitation, broadcasting, reproduction on microfilms or in any other physical way, and transmission or information storage and retrieval, electronic adaptation, computer software, or by similar or dissimilar methodology now known or hereafter developed.
The use of general descriptive names, registered names, trademarks, service marks, etc. in this publication does not imply, even in the absence of a specific statement, that such names are exempt from the relevant protective laws and regulations and therefore free for general use.
The publisher, the authors and the editors are safe to assume that the advice and information in this book are believed to be true and accurate at the date of publication. Neither the publisher nor the authors or the editors give a warranty, express or implied, with respect to the material contained herein or for any errors or omissions that may have been made.

Printed on acid-free paper

Springer-Verlag GmbH Berlin Heidelberg is part of Springer Science+Business Media
(www.springer.com)

Preface

The 24th International Conference on Database and Expert Systems Applications (DEXA 2013), with proceedings published as volumes 8055 and 8056 in Springer's Lecture Notes in Computer Science, featured some outstanding keynote presentations and regular articles. As with previous editions of the DEXA conference, the Program Co-chairs of DEXA 2013 invited some of the authors to submit extended papers to a special issue of the Springer journal Transactions on Large-Scale Data- and Knowledge-Centered Systems (TLDKS). Following these invitations, both keynote papers and eight regular articles were submitted. Apart from the keynotes, each submission was carefully assessed by at least two (often more) recognized experts in the respective field. In total, 35 reviews were received, most of them of excellent quality. After two rounds of revisions, five of the eight regular papers were accepted for inclusion in this special issue, in addition to the two keynote papers.

The contributions in this special issue address a range of important modern subject areas in data-centric systems and applications, inclusive of argumentation, e-government, business processes, predictive traffic estimation, semantic model integration, top-k query processing, uncertainty handling, graph comparison, community detection, genetic programming, and web services. In the DEXA tradition, all contributions distinguish themselves by the novelty and innovation they bring to these subject areas.

The first keynote paper is authored by the presenter, Trevor Bench-Capon, from the University of Liverpool, England, together with his colleagues Katie Atkinson, also from Liverpool, and Adam Wyner, affiliated with the University of Aberdeen in Scotland. Each of them is a distinguished expert in the field of *computational argumentation*. Theoretical work on argumentation usually focuses on inferring consistent sets of facts, rules, and assumptions that support each other and form coherent positions on an issue. In addition to that, the authors investigate an argumentative form of *practical reasoning*, for justifying decisions about actions, as opposed to *theoretical reasoning*, which merely deals with what is the case. The particular application of practical reasoning investigated in the keynote paper entitled "Using Argumentation to Structure E-Participation in Policy Making" is the engagement of citizens in dialogues with governmental entities about policies, by means of electronic computational devices using argumentation.

For the second keynote paper, several authors from the Software Competence Center in Hagenberg and the close-by University of Linz, both in Austria, have collaborated under the leadership of the original keynote presenter, Klaus-Dieter Schewe. Traditionally, the many different aspects of business process modeling have been addressed by different models. This makes it nearly impossible for stakeholders to develop sufficient levels of trust in the quality of the business processes as a whole, preventing mission-critical analysis and decision making. In their contribution "Horizontal Business Process Model Integration," the authors propose the integration of different process models by specifying their semantics uniformly with abstract state machines.

The proposal is driven by the strong desire to derive targeted levels of quality on the business processes in their entirety, which can be derived by taking advantage of the rigorous verification and formal validation techniques that are a built-in feature of abstract state machines.

The paper entitled "Exact and Approximate Generic Multi-criteria Top-k Query Processing" is concerned with the ranking of answers to queries. It is authored by Mehdi Badr and Dan Vodislav, both from the University of Cergy-Pontoise, France. Top-k queries ask for the k best answers, where answer goodness is ranked according to the scores produced by criteria that are stated in the query as *ranking predicates*. Most top-k query processing algorithms are tailored to work for a specific kind of access to ranking predicate scores, which may either be sorted, or at random, or both sorted and random. An important contribution of the work by Badr and Vodislav is that they propose a framework for *generic* top-k processing, in which it is possible to express and analyze any top-k algorithm, regardless of whether it uses some strict (i.e., either sorted or random) or some hybrid form of access. They have also elaborated on extended, more generic variants of previously proposed algorithms, such that they become easily comparable. Many existing approaches for top-k query processing only compute exact results. While, in principle, exactness is desirable, it all too often comes at the expense of execution time, so that more efficient approximations have to be resorted to. The generic framework presented in this paper results in a thorough performance analysis and comparison of exact and approximate top-k algorithms.

A pragmatic and highly interesting special-purpose solution to the problem of timely traffic route prediction is proposed in the article on "Continuous Predictive Line Queries for On-the-Go Traffic Estimation," authored by Lasanthi Heendaliya, Dan Lin, and Ali Hurson from the Missouri University of Science and Technology in Rolla, Missouri, USA. Instead of simply offering predictions of optimal routings, called *lines*, that are based on static snapshots of traffic conditions, the paper proposes a new type of spatial-temporal queries, called *continuous predictive line queries*. These result in continuously monitoring traffic dynamics, and return adjusted route suggestions whenever the monitored circumstances change significantly. Thus, a much more up-to-date feedback loop to users on the road is enabled. Details about a novel data structure and the pseudo-code of algorithms for implementing the proposed approach are provided in the paper, as well as assiduous analyses of its performance and costs. The evaluations reveal quantifications of efficiency and effectiveness that improve conventional static predictions.

The recent resurgence of graph and network data types in the framework of graph databases is reflected in the paper entitled "Query Operators for Comparing Uncertain Graphs," authored by a team of researchers from Georgetown University, Washington DC, USA, consisting of Denis Dimitrov, Lisa Singh, and Janet Mann. Graph comparison is useful for detecting deviations and for hypothesizing properties of networked structures by analogy from known properties of similar networks. Several query languages feature operators for comparing graphs and subgraphs. Others have proposed extensions of graph models by incorporating vague attribute values, as well as uncertainties about the presence or absence of vertices and edges. However, the combination of comparison and uncertainty as presented in this paper is innovative. Dolphin observation and citation networks are two showcases used to illustrate the query language and its ability to analyze real-world uncertain graph data.

A performance study shows the viability of the proposed framework for reasonably large graphs.

The paper entitled "Fast Disjoint and Overlapping Community Detection" is authored by Yi Song and Stéphane Bressan, from the University of Singapore, and Gillian Dobbie, from the University of Auckland, New Zealand. Grosso modo, their work falls into the topic area of *social networks*. Communities are defined as the subgraphs of such networks that feature a significantly high interconnectivity among their members. The detection of such communities is useful in many applications, such as sociology, biology, marketing, health care, etc. Many approaches to community detection focus on partitions of disjoint communities (which, for example, is natural for distributed networks of data stores that are fragmented by some node failures or broken connections). However, the algorithms presented in this paper can be parallelized for scaling them up to possibly large networks with overlapping communities. Such overlaps are typical for social networks and critical applications such as epidemics control or, more generally, networks with nonlocal interconnectivity. The metrics used for empirically quantifying the effectiveness and runtime efficiency of the algorithms involve network dimensions such as modularity, conductance, internal density, cut ratio, community size, and weighted community clusters. The measurements of effectiveness and efficiency also serve to compare the algorithms with state-of-the-art solutions, with favorable results for the approach presented by the authors.

Finally, the paper "A Hybrid Approach using Genetic Programming and Greedy Search for QoS-Aware Web Service Composition," by Hui Ma, Anqi Wang, and Mengjie Zhang from the Victoria University of Wellington, New Zealand, offers insights into synergies obtained by applying methods from the hybrid fields of genetic programming and greedy search, resulting in surprising improvements in web service composition. The difficulties of the latter have grown proportionally to a tremendous increase of web services in recent years. The authors propose the use of a greedy algorithm for generating populations of candidate services, on which genetic-pro-gramming-based mutations are performed in order to obtain optimized service com-positions. The validity of the proposal is made plausible by an experimental study based on public benchmark test cases with repositories of large quantities of web services and pertinent properties. Moreover, the authors elaborate on an extension of their approach in terms of optimizing solutions according to some given quality of service criteria.

We would like to thank all authors for their contributions to this special issue. We are grateful to all reviewers for their invaluable work in reviewing the papers and ensuring the high quality of this collection of articles. Last, but not least, our gratitude goes to Gabriela Wagner, whose editorial assistance and handling of all the commu-nication with the authors and the reviewers finally made this volume possible.

December 2014

Hendrik Decker
Lenka Lhotska
Sebastian Link

Organization

Editorial Board

Reza Akbarinia	Inria, France
Bernd Amann	LIP6 – UPMC, France
Dagmar Auer	FAW, Austria
Stéphane Bressan	National University of Singapore, Singapore
Francesco Buccafurri	Università Mediterranea di Reggio Calabria, Italy
Qiming Chen	HP Labs, USA
Tommaso Di Noia	Politecnico di Bari, Italy
Dirk Draheim	University of Innsbruck, Austria
Johann Eder	Alpen Adria Universität Klagenfurt, Austria
Stefan Fenz	Vienna University of Technology, Austria
Georg Gottlob	Oxford University, UK
Anastasios Gounaris	Aristotle University of Thessaloniki, Greece
Theo Härder	Technical University of Kaiserslautern, Germany
Andreas Herzig	IRIT, Paul Sabatier University, France
Hilda Kosorus	FAW, Austria
Dieter Kranzlmüller	Ludwig-Maximilians-Universität München, Germany
Philippe Lamarre	INSA Lyon, France
Lenka Lhotská	Technical University of Prague, Czech Republic
Vladimir Marik	Technical University of Prague, Czech Republic
Mukesh Mohania	IBM Research, India
Franck Morvan	IRIT, Paul Sabatier University, France
Kjetil Nørvåg	Norwegian University of Science and Technology, Norway
Gultekin Ozsoyoglu	Case Western Reserve University, USA
Themis Palpanas	Paris Descartes University, France
Torben Bach Pedersen	Aalborg University, Denmark
Günther Pernul	University of Regensburg, Germany
Klaus-Dieter Schewe	Johannes Kepler University Linz, Austria
David Taniar	Monash University, Australia
A Min Tjoa	Vienna University of Technology, Austria
Chao Wang	Oak Ridge National Laboratory, USA

External Reviewers

Eva Alfaro	Instituto Tecnológico de Informática, Universidad Politécnica de Valencia, Spain
José Enrique Armendáriz-Iñigo	Universidad Pública de Navarra, Spain
Jan Chomicki	University at Buffalo, USA
Deborah Dahl	Conversational Technologies, USA
Samira Daruki	University of Utah, USA
Gill Dobbie	University of Auckland, New Zealand
Sven Groppe	Lübeck University, Germany
Oren Halvani	Fraunhofer Institute, Germany
Ibrahim Hamidah	Universiti Putra, Malaysia
Lipyeow Lim	University of Hawaii at Manoa, USA
Chuan-Ming Liu	National Taipei University of Technology, Taiwan
Lin Liu	Kent State University, USA
Swarup Kumar Mitra	MCKV Institute of Engineering, India
Francesc Munoz-Escoi	Polytechnic University of Valencia, Spain
Nam P. Nguyen	Towson University, USA
Odysseas Papapetrou	SoftNet Lab, Technical University of Crete, Greece
Rodolfo F. Resende	Federal University of Minas Gerais, Brazil
Klaus-Dieter Schewe	Software Competence Center Hagenberg, Austria
Bala Srinivasan	Monash University, Australia
Efstathios Stamatatos	University of the Aegean, Greece
Marco Vieira	University of Coimbra, Portugal
Maksims Volkovs	University of Toronto, Canada
Ye Yuan	Northeastern University, China
Zhong-Yuan Zhang	Central University of Finance and Economics, China
Roger Zimmermann	National University of Singapore, Singapore

Contents

Using Argumentation to Structure
E-Participation in Policy Making

Trevor Bench-Capon[1](✉), Katie Atkinson[1], and Adam Wyner[2]

[1] Department of Computer Science, University of Liverpool, Liverpool, UK
tbc@liverpool.ac.uk
[2] Department of Computing Science, University of Aberdeen, Aberdeen, UK

Abstract. Tools for e-participation are becoming increasingly important. In this paper we argue that existing tools exhibit a number of limitations, and that these can be addressed by basing tools on developments in the field of computational argumentation. After discussing the limitations, we present an argumentation scheme which can be used to justify policy proposals, and a way of modelling the domain so that arguments using this scheme and attacks upon them can be automatically generated. We then present two prototype tools: one to present justifications and receive criticism, and the other to elicit justifications of user-proposed policies and critique them. We use a running example of a genuine policy debate to illustrate the various aspects.

Keywords: E-participation · Argumentation · Dialogues · Deliberation · Values · Policy making

1 Introduction

An important feature of democracies is that citizens can engage their governments in dialogues about policies. Traditionally this was done by writing letters: government departments employed a large number of people whose main function was to reply to these letters on behalf on the Ministers to whom the letters were addressed[1]. Although a large number of letters concerned the particular individual circumstances of the writer, others were directed towards general policy matters. Such letters tended to fall into one of three types: some were in pursuit of information and sought a justification of some policy or action; some (probably the most common) objected to all or some aspects of a policy; a third type made policy proposals of their own. The policies we have in mind have a very broad range, running from particular local issues with a small impact to issues of national importance which potentially impact on all citizens. In this introduction we will characterise each of the three different types of engagement

---

[1] The first author worked as a Civil Servant for the UK Department of Health and Social Security in the late seventies, and part of his duties was replying to such correspondence.

© Springer-Verlag Berlin Heidelberg 2015
A. Hameurlain et al. (Eds.): TLDKS XVIII, LNCS 8980, pp. 1–29, 2015.
DOI: 10.1007/978-3-662-46485-4_1

and illustrate them with reference to a simple motivating example concerning a proposal by a local council to close a community library. In subsequent sections we will use the introduction of cameras by a national government to detect motorists who are exceeding the speed limits as a running example which we will model and discuss in detail.

For the first type of letter described above, the reply need only state a justification, which could be a stock reply: once a justification has been developed it can be sent in response to all such inquiries. Relating this to our example scenario, when the proposal to close the library is announced people may seek information about the number of users and trends in usage over a time period, as well as information about the running costs, both of which are likely to form part of the justification for the proposed closure. The response is also likely give reasons for closure in terms of usage or budget, or some other motivation.

For the second type of letter, a justification is not enough: to produce a satisfactory reply the respondent needs first to understand what the citizen objects to, and then to give an answer to the specific points. This may not be entirely straightforward: often the writer will be unclear or ambiguous or lack focus. In our example, the citizen might object to the closure of the library on a number of different grounds such as a lack of alternative libraries in the local area, or the council's allocation of funds across its services, which must be first disentangled and then answered separately.

For the third type of letter, even more is required. First a well formulated proposal must be stated, and then that proposal can then be critiqued from the standpoint of the government's own beliefs and values. Both of these might prove difficult. Formulating a policy is not an easy task, and so some considerable effort might be needed to get the proposal into a coherent form. Also the critique might require a variety of different kinds of knowledge, ranging from facts, through economic models and budgetary constraints, to value choices. In our example scenario, an alternative action that might be proposed is the creation of a mobile library to serve a number of different communities and save costs.

A valuable by-product of this correspondence was that it enabled the mood of the public to be gauged: those receiving and replying to this correspondence could get a feel for which aspects of policy were popular and which were unpopular, and which alternatives were well supported. But such knowledge tended to be anecdotal and impressionistic: the paper process did not lend itself to systematic quantification.

Nowadays e-mail and the internet offer a better way of conducting this kind of dialogue. But while communication is quicker and more convenient, the task remains difficult. It is still hard to formulate policy proposals, justifications and critiques cogently. Nor does standard e-mail correspondence lend itself to statistical aggregation. But there is no obligation simply to replicate the existing process. E-participation does offer opportunities to provide support for understanding inquiries, formulating replies and the aggregation task required to make sense of the feedback. Unfortunately these opportunities have rarely been taken.

Current e-participation systems too often lack structure. Most commonly they take the form of petitions or threaded discussions. Petitions allow the expression of general feelings, but they are unable to express objections with precision. Too often they are ill expressed and conflate a variety of different arguments, so that it is not clear what people are subscribing too. Threaded discussions allow people to feel that they have expressed their views, but they too lack structure. Thus arguments are typically ill-formed, and the lack of structure also makes comparison, aggregation and assimilation difficult. In consequence, government replies are often general, bland and superficial; they fail to address the particular objections of the citizens; and the views expressed by the citizens remain hard to quantify. To address these issues, we believe that tools that are firmly grounded on a well defined model of argument are needed.

Following a discussion of some existing tools, and their limitations, we introduce our model of argument. We present an underlying semantic structure and argument scheme for the justification of policy proposals, along with ways of critiquing such justifications in terms of its structure. We also offer a detailed example, instantiating the formal structure with a representation of a real policy debate: whether speed cameras should be introduced on major roads. This will form a running example for use in the following sections where we introduce our two tools, directed in turn at each of the second and third tasks described above, and at the collection and aggregation of information from the dialogues. Finally we offer some concluding remarks.

2 Existing Tools

From a developer's point of view, a key consideration in designing and building on-line tools for e-participation is the trade-off between the amount of structure provided by the tool and its ease of learning and use. Since the target audience is the general public, participation must be fostered by making the interactive system as straightforward to use as possible. If, however, the responses are to be meaningfully analysed in terms of their content, then considerable structure needs to be imposed on the data. In this section we will discuss some existing tools[2] and then summarise what we see as their limitations.

2.1 E-Petitions

The simplest e-participation tool is the e-petition. This allows people to register a petition, criticising a policy or advocating a change of policy, and provides the means for other people to endorse it. This is the modern version of a very traditional method of expressing grievances: since at least the eighteenth century

---

[2] The IMPACT project ran from January 2010 until December 2012. The tools described here are predominately those that provided the context for the developments of that project, which are the main topic of this paper. Since then, social media, especially Twitter, has become widely used, and several e-participation developments have attempted to reflect this. Thus the focus remains very much on the communications channel, and it remains true that there has been little attention paid to providing more structure and coherence to the utterances.

people went round with paper petitions gathering signatures and presenting them to their rulers. Given enough signatures, the government may issue a reply, or the issue might even be debated in Parliament. But really, apart from convenience, the e-petition represents little by way of progress from the paper version.

Under the previous Labour administration (1997–2010), the UK government introduced a much used e-petition site[3]. The motivation was stated on the site as *e-petitions is an easy way for you to influence government policy in the UK*. These e-petitions could address anything for which the government is responsible. Once a petition got at least 100,000 signatures, it was eligible for debate in the UK parliament. A similar site was also used by the US government where an official response was issued once the petition reached a threshold number of signatures.

Whilst these e-petitions indeed proved easy to use, easy to respond to and facilitated signature collection (one particular petition in the UK gained over 1.81 million electronic signatures), the *quality* of engagement they offered is questionable. Such e-petitions are simply electronic versions of paper petitions, and they suffer from the same shortcomings as paper versions, the most significant being the conflation of a number of issues into one catch-all statement. As Dr. Samuel Johnson wrote back in the eighteenth century:

> The petition is then handed from town to town, and from house to house; and, wherever it comes, the inhabitants flock together, that they may see that which must be sent to the king. Names are easily collected. One man signs, because he hates the papists; another, because he has vowed destruction to the turnpikes; one, because it will vex the parson; another, because he owes his landlord nothing; one, because he is rich; another, because he is poor; one, to show that he is not afraid; and another, to show that he can write.

The recipient of the petition can only assume that by signing, the signatory agrees wholeheartedly with all of the (potentially) multiple points raised in the statement. This makes it easy to over simplify and to blur the issues since it is likely that individuals object for different reasons. Consider, for example, one of the most popular petitions on the UK site which criticised a proposed reduction in the UK national speed limit on roads. The petition objected that the reduction would not make a difference to road deaths and that the subsequent cut in carbon emissions would be too insignificant to justify the speed limit reduction. Signing such a petition is an 'all-or-nothing' statement with no room to discriminate between (or even acknowledge) the two very different objections raised. In a word, the petitions lack structure. The responses provided by the government were also at a general level and not able to recognise or address particular concerns, and so typically failed to satisfy anyone fully. We need the opinions to be presented in a coherent, well reasoned, structure: *arguments* rather than mere *assertions*.

---

[3] A very similar site, launched by the current Conservative administration, is currently (2014) available at http://epetitions.direct.gov.uk.

2.2 Free Text Based Tools

There have been several proposals for policy-making support tools in the European Union and the United States which use currently available wiki, comment, email, or social networking technologies (see [9,16] for discussion of other tools such as IBIS+, Compendium, DebateGraph). We discuss several of these briefly in order to set the context for the contribution of our Structured Consultation Tool (which we refer to herein as the SCT).[4]

The United Kingdom's Cabinet Office *Public Reading* website[5], presented the Protection of Freedoms Bill, using a website that unfolds the proposed bill, allowing on-line readers to look at specific sections. At the bottom level, the user can use a threaded comment facility to respond to a particular portion or responses made by other users. With the Public Reading tool, it is difficult to get an overview understanding of the whole policy and the relation of responses to it. Thus, the role and impact of responses is not highlighted. There is no support for analysing the responses, which is then done "manually" by analysts of the consultation, making the contribution of the responses to any subsequent development of the policy draft obscure. Moreover, while the responses are specifically linked to parts of the legislation, the unconstrained nature of the responses means the consultation is unstructured and unsystematic. Not only does this allow inappropriate or irrelevant responses, but it may not elicit the kind of important or useful information that is the primary motivation for the consultation in the first place. The Bill itself proposes a solution to some legislative problem; comments on the Bill may discuss alternative solutions. Yet understanding the Bill or alternative solutions may rest on the motivations and justifications underlying the solutions, for example, in terms of social values that the solution promotes. Making these motivations and justifications overt would further support rational analysis and understanding of the Bill, which in turn would better represent the stakeholders' interests and objectives.

Like the UK Prime Minister's Office *e-petition* site discussed above, the European Commission's *The European Citizens' Initiative* facilities allow citizens to electronically create, sign, and submit petitions.[6] By the same token, these tools can be used to "vote" on a policy proposal. The tools, which enable respondents to submit petitions, are web-based versions of what is has been traditionally accomplished manually. Both of these tools contribute to the policy formulation stage of the policy-making cycle, but not to the comment stage. There is no analytic framework. A particular problem is that it is unclear exactly *what* respondents are signatory to; that is, it provides an unrefined *all or nothing* representation of a point of view, whereas there may well be respondents who agree with some parts of the proposal, but not other parts, yet nonetheless sign on to the whole. What is needed is support to *differentiate* and *draw out* such subtle alternative viewpoints.

---

[4] All websites accessed April 24, 2014.

[5] http://publicreadingstage.cabinetoffice.gov.uk/ (archive only).

[6] http://epetitions.direct.gov.uk/.
   http://ec.europa.eu/citizens-initiative/public/welcome.

Other initiatives aim to improve the quality of comments on proposed legislation. The US General Services Administration used a tool to support consultation, *ExpertNet*, which drew upon "crowdsourced" expertise and attempted to structure responses with social networking facilities such as ranking responses, providing specific questions for community voting and annotating responses, among others. While this does give indicative information on respondents' reactions, the legislation is not represented in an analytic form, let alone a form able to support machine analysis. Rather, although the content of the legislation and the reactions to it must be further analysed, there is no analytic framework to support this. There are additional issues raised about how to identify, certify, and monitor the community of experts. The *RegulationRoom* is an academically hosted facility for commenting on proposed legislation, providing guidelines on effective comments. This is more substantive than ExpertNet, but it requires highly skilled individuals to follow the guidelines; it may best suit respondents who already participate in policy consultations rather than untrained members of the public.[7]

Finally, in the US state of Massachusetts, legislators provide a wiki tool, *Lex-Pop*, to "crowdsource" the incremental development of legislation.[8] The question here concerns who is in a position to use such a tool, not just in terms of representing the interests of others and reasoning about legislation, which often requires a deep understanding of law and how to author legislation, but also reasoning about legal values and consequences. The success of current wikis (e.g. Wikipedia) rests on an often small coterie of self-selected, self-regulating authors who write about specialist topics, where questions and controversies can be left unresolved and where there are no legislated consequences.

Despite these drawbacks, these past and current tools and initiatives are clearly potentially important and useful in leveraging current technologies to draw in greater citizen participation to policy-making by making participation easier and improving the informativeness of feedback. However, providing the means to address or avoid these limitations would positively impact on policy making. In particular, the tools discussed above do not further the substantive semantic analysis of the comments in a form that supports machine-processing of rich, complex information, particularly where the comments introduce conflicts and inconsistencies that must be reasoned with. That is, they do not make use of current thinking or techniques found in Artificial Intelligence on argumentation.

2.3   Structured Tools

A key issue we have raised here with respect to tools that solicit user input in free text is how and where to impose structure to identify the arguments proposed so that the analysis of the opinions can be made meaningful, and even supported through computational analysis. An alternative, relatively untried in

---

[7] http://expertnet.wikispaces.com/.
http://regulationroom.org/.

[8] http://lexpop.org/.

practice, where success seems to be judged by the quantity rather than the quality of responses received, is to oblige users to conform to a restrictive structure. This may, however, inhibit their interaction or require them to understand the underlying theory. Users may then make mistakes, and their responses be precise but wrong, which is even worse than being vague. Despite the difficulties, a number of research systems have been developed with the intention of providing a better level of support. We briefly discuss some of the better known and the issues they raise.

One category of tool is argument mapping tools. Araucaria [19] is one example which enables users to mark up the premises and conclusions of arguments, and indicate particular argumentation schemes identifying patterns of reasoning. Whilst the mark-up requires users to think more deeply about the structure of their arguments, there still remains no guarantee that the semantics of the marked up text is coherent and consistent since users simply decide what text to label as premises and conclusions and what the inferences are. In consequence the tool can accept invalid mark-ups and typically there are several different, equally valid, mark-ups.

Other on-line argument mapping tools include *Debatepedia* and its replacement, *Debatabase*[9]. These are on-line 'wikis' containing an ever growing collection of arguments and debates within which users can express pros and cons of a range of issues. Although democratic in that users can freely modify others' contributions, the arguments entered are not required to conform to any particular semantics that would support coherence and argument evaluation, and so it is often difficult to relate the various points made, and to evaluate the status of the debate.

Still more structure is imposed by systems that have been built using the IBIS (Issue Based Information Systems) model of argument [15]. IBIS enables a particular problem or issue to be decomposed into a number of different positions. Arguments can then be created to attack or defend the positions until the issue is settled (possibly by a vote). A collaborative decision support system that uses this model is HERMES [14] (as does its predecessor Zeno [12]). Evaluation showed that although users enjoyed using the system it was not easy to learn and difficulties were experienced understanding the argumentation content of the system, casting doubt on the usefulness of its output.

More recently there has been a shift towards the development of tools that make use of ideas and trends from social media. A comprehensive survey of the state-of-the-art in web-based argumentation tools, which also covers a number of the tools we have mentioned above, can be found in [20].

2.4   Limitations of Existing Tools

In this section we will summarise the limitations of existing tools, which we hope to address using the model-based tools we will describe in later sections. The first problem relates to the analysis of the responses. The current tools have a

---

[9] http://idebate.org/debatabase.

focus on usability and accessibility, and are indeed easy and convenient to use. In consequence they have proved highly popular and successful in attracting participants. The downside, however, is that there is a lot of unstructured data collected: too much data to be able to use these responses to inform policy making. The current tools allow people to express their opinions, but do not enable these opinions to feed easily into policy making. This raises the following questions:

- How can we systematically organise the analysis of comments?
- How can we organise the information to accurately identify issues and consult participants in further depth?
- The abundance of claims, counter-claims, evidence, points of view, etc. results in a rich 'web' of information. How can we manage so large a quantity of heterogeneous data, and reason effectively with it?

As well as the quantity of data, the fact that it is unstructured - typically simply free text authored by non specialists - presents problems:

- Comments are in an unstructured and unsystematic format. While threaded lists are often used, enabling people to follow and continue a discussion to some extent, it remains difficult even for a skilled human, let alone a machine, to extract meaningful information in any systematic way.
- Threaded lists can often wander away from their original topic, so that they may become irrelevant, or relevant information may appear under unrelated headings.
- Comments are not sufficiently fine-grained to be as informative as may be needed if they are to impact on policy making. Underlying motivations and justifications are often insufficiently specific and are also often left implicit or taken for granted by users unused to framing their opinions for a general audience.

Third, since the focus is on allowing people to "have their say", many of the contributions are ill informed, biased and unbalanced. But the bias may also come from the analyst: since there is simply a mass of unstructured information, it is possible to cherry-pick the comments that one will make use of. Experts who mediate, analyse, and summarise the comments can bias information or obscure the relation between comments and policy outcomes. Outlier, hybrid, challenging, and novel positions on issues may get 'lost'. Thus the process does not produce objective, transparent results.

Finally we can see problems with the model of interaction itself. The task that the participants are asked to perform is really rather difficult, both for the citizens and the officials. They are being asked to:

- construct a coherent argument
- maintain relevance and focus
- get the facts right
- understand an argument

- identify and answer specific objections
- answer at the correct level of detail
- relate, combine and aggregate arguments.

This adds up to a rather demanding skill set which we would not expect everyone to possess. But current tools exhibit:

- Lack of support for reasoning processes (inference, modelling, consistency, alternative policy positions).
- Little interaction and feedback among stakeholders and between stakeholders and the consultative body. There is no deliberation.

Given all the issues we have raised, we see a clear need for on-line opinion gathering tools to be grounded on some solid semantic foundation whilst retaining their usability. To achieve this, we look to multi-agent systems, and in particular how the reasoning of the agents in a system can be supported by a computational model of argument. In the next section we describe an approach from this field that can provide the backbone of support for tools that can be used to improve on-line opinion gathering.

3 Policy as Practical Reasoning

While current systems make excellent use of the available technology, they serve mainly as a communications channel and lack the domain expertise and knowledge, which would be required to provide the users with support in formulating and structuring their contributions and to facilitate understanding and analysis. We therefore look to computational argumentation to overcome these deficiencies. Computational argumentation provides us with methods of argument representation and evaluation. This provides the expertise, although when building tools to support citizen participation, we must not neglect to strike a balance between the use of structured argument and ease of use of the tools. But computational argumentation also requires domain knowledge to instantiate the argumentation structures, and so we need an underlying model of the domain as well as a model of argumentation.

In this section we will describe how the model of argumentation based on argumentation schemes as proposed in [22], and in particular the argumentation scheme for practical reasoning proposed in [4], can supply the model of argumentation, and how Action-based Alternating Transition Systems (AATS), developed in multi-agent systems for reasoning about joint actions and coalitions [24] provide an appropriate model with which to store domain knowledge. Specifically we will base our tools on [2] which used AATSs to provide a formal basis for the practical reasoning argumentation scheme of [4].

3.1 Argumentation Scheme for Practical Reasoning

Practical reasoning is used to justify, or argue for, decisions as to what to *do* (in contrast to theoretical reasoning which concerns what is the case). As such we need to recognise that different people may decide, justifiably, to do different things, because they have different desires, aspirations and preferences. As Searle [21] puts it, whereas in theoretical reasoning we attempt to fit our beliefs to the world, in practical reasoning we try to fit the world to our desires: and our desires differ.

Normally there will be aspects of the current state that the agent likes, and aspects that it does not like. So, with respect to change, the agent will have four possible motivations:

- To make something currently false true (*achievement goal*).
- To make something currently true false (*remedy goal*).
- To keep something true true (*maintenance goal*).
- To keep something false false (*avoidance goal*).

What an agent wants can be specified at several levels of abstraction. Suppose an agent enters a bar on a hot day and is asked what it wants. The agent may reply:

- I want to increase my happiness.
- I want to slake my thirst.
- I want a pint of lager.

The first reply relates to something which is almost always true, and for the sake of which other things are done. Normally there will be several things that promote this state. The second is a specific way of increasing happiness: it is a remedy goal. There is an element of the current situation the rectification of which would increase the happiness of the agent. Again there are several ways of bringing this about. Finally the third reply identifies a specific way of remedying the situation: the agent selected a lager in preference to water, juice, etc. It is a specific condition under which the goal will be satisfied. Previous work such as [2] has used *values*, *goals* and *circumstances* to refer to these three levels of abstraction. In [2] these levels are related to motivate or justify a choice through expression as an argument scheme. Argument schemes provide templates to capture stereotypical patterns of reasoning and they have associated with them critical questions to probe the presumptive conclusions that can be drawn by instantiating the schemes. A variety of different schemes is documented in the informal logic literature [22], and they are increasingly being used in computational argumentation. The following argument scheme for practical reasoning distinguishes the three levels of abstraction discussed above:

PRAS: In the current circumstances R, I should perform action A, to bring about new circumstances S, which will achieve goal G and promote value V.

Applied to the example above, this would give: *In the pub (current circumstances), I should order a lager (action), to have a drink (new circumstances), which will slake my thirst (goal), which will increase my happiness (value).* Policy making can be seen as conforming to this model. The policy makers will have some values which they wish to pursue. This language of values is very common in contemporary politics, and voters often choose between parties on the basis of their perceived values rather than on the basis of specific policy proposals. Values can only be realised, however, through concrete action, and this requires a set of goals to be adopted. Finally ways of realising these goals must be identified and actions to bring the required circumstances about must be identified. Thus we can see policy making as a form of practical reasoning, and the argumentation scheme of [2] as a form of argument for policy justification.

We shall next describe the underlying semantic structure, the AATS, extended to include values, and how this structure can be used to instantiate arguments of the form of PRAS. We then consider how such arguments can be attacked and defended.

3.2   A Semantic Structure for Practical Reasoning

Action-Based Alternating Transition Systems (AATSs) were originally presented in [24] as semantical structures for modelling game-like, dynamic, multi-agent systems in which the agents can perform actions in order to modify and attempt to control the system in some way. These structures are thus well suited to serve as the basis for the representation of arguments about which action to take in situations where the outcome may be affected by the actions of other agents. First we recapitulate the definition of the components of an AATS given in [24].

Defnition 1: AATS An *Action-based Alternating Transition System* (AATS) is an $(n + 7)$-tuple $S = \langle Q, q_0, Ag, Ac_1, \ldots, Ac_n, \rho, \tau, \Phi, \pi \rangle$, where:

- Q is a finite, non-empty set of *states*;
- $q_0 \in Q$ is the *initial state*;
- $Ag = \{1,\ldots,n\}$ is a finite, non-empty set of *agents*;
- Ac_i is a finite, non-empty set of actions, for each $i \in Ag$ where $Ac_i \cap Ac_j = \emptyset$ for all $i \neq j \in Ag$;
- $\rho : Ac_{Ag} \to 2^Q$ is an *action pre-condition function*, which for each action $\alpha \in Ac_{Ag}$ defines the set of states $\rho(\alpha)$ from which α may be executed;
- $\tau : Q \times J_{Ag} \to Q$ is a partial *system transition function*, which defines the state $\tau(q, j)$ that would result by the performance of j from state q – note that, as this function is partial, not all joint actions are possible in all states (cf. the pre-condition function above);
- Φ is a finite, non-empty set of *atomic propositions*; and
- $\pi : Q \to 2^\Phi$ is an interpretation function, which gives the set of primitive propositions satisfied in each state: if $p \in \pi(q)$, then this means that the propositional variable p is satisfied (equivalently, true) in state q.

AATSs are particularly concerned with the joint actions of the set of agents Ag. j_{Ag} is the joint action of the set of n agents that make up Ag, and is a tuple $\langle \alpha_1,\ldots,\alpha_n \rangle$, where for each α_j (where $j \leq n$) there is some $i \in Ag$ such that $\alpha_j \in Ac_i$. Moreover, there are no two different actions α_j and $\alpha_{j'}$ in j_{Ag} that belong to the same Ac_i. The set of all joint actions for the set of agents Ag is denoted by J_{Ag}, so $J_{Ag} = \prod_{i \in Ag} Ac_i$. Given an element j of J_{Ag} and an agent $i \in Ag$, i's action in j is denoted by j^i.

To represent the values within our reasoning framework, the AATS structure must be extended to enable the representation of values, which was done in [2]. For this, a set V of values was introduced, along with a function δ to enable every transition between two states to be labelled as either promoting, demoting, or being neutral with respect to each value.

Definition 2: AATS+V

Given an AATS, an AATS+V is defined as follows:

- V is a finite, non-empty set of values.
- $\delta : Q \times Q \times V \rightarrow \{+, -, =\}$ is a *valuation function* which defines the status (promoted (+), demoted (–) or neutral (=)) of a value $v_u \in V$ ascribed to the transition between two states: $\delta(q_x, q_y, v_u)$ labels the transition between q_x and q_y with one of $\{+, -, =\}$ with respect to the value $v_u \in V$.

An *Action-based Alternating Transition System with Values* (AATS+V) is thus defined as a $(n + 9)$ tuple $S = \langle Q, q_0, Ag, Ac_1, ..., Ac_n, \rho, \tau, \Phi, \pi, V, \delta \rangle$.

This formalism was used in [2] to formalise the PRAS argumentation scheme introduced informally in the previous section.

Definition 3: PRAS

In the initial state $q_0 = q_x \in Q$,
Agent $i \in Ag$ should participate in joint action $j_n \in J_{Ag}$

where $j_n^i = \alpha_i$,

and $\tau(q_x, j_n)$ is q_y,

and $p_a \in \pi(q_y)$ and $p_a \notin \pi(q_x)$, or $p_a \notin \pi(q_y)$ and $p_a \in \pi(q_x)$,

and for some $v_u \in V$, $\delta(q_x, q_y, v_u)$ is +.

3.3 Attacking and Justifying Policy Arguments

An important feature of argumentation schemes as described by Walton [22] is that they only *presumptively* justify their conclusions. Moreover, each argumentation scheme has its own characteristic ways of being attacked. Walton termed these methods of attack "critical questions". In [2] seventeen ways to attack arguments based on PRAS were identified, and these were divided into three different types of critical question:

- *problem formulation*: deciding what the propositions and values relevant to the particular situation are, and constructing the AATS. There are eight such attacks. These concern the propositions used in the state descriptions, the actions available and their effects, which values exist and which transitions promote and demote them.

– *epistemic reasoning*: determining the initial state in the structure formed at the previous stage, and which joint action will be performed. There are two such attacks: one challenging the current circumstances and one questioning the anticipated behaviour of the other agents involved in joint actions.
– *choice of action*: These are the remaining seven attacks, which involve consideration of alternative ways of achieving goals and values; side effects that that will demote values, and passing up an opportunity to promote some other value. Essentially these will be resolved according to the value preferences of the individual acting as the audience for the argument.

These different categories of attack can be seen as requiring resolution at different levels. The problem formulation attacks are the most fundamental: they express differences about what is relevant, the results of actions, what promotes values, and the like. Differences at this level necessitate different models of the world: they require those disagreeing to have a different AATS in mind. Epistemic questions are not fundamental: they do not need a different AATS, but they do seek to establish agreement as to where we are in the structure and what paths we will follow. Either there must be a recognised way of determining the "truth of the matter", or else disputants must agree to proceed on the basis of assumptions. For choice of action, however, disagreement is to be expected, since differences turn on the different priorities given to the various social values involved. The set of arguments generated will be the same, but they will be evaluated differently by different audiences.

With regard to answering such attacks, analysis in [9] suggested that the underlying claims were typically justified by citing some credible source. This might be expert opinion, the conclusion of some scientific survey, a public opinion poll, witness testimony, or - in the case of value statements - party manifestos. Rarely were arguments from first principles used. Credible source arguments cover a number of the argument schemes given in [22]. The work of [9] suggests a three ply model: a policy justification using PRAS; a challenge based on one of the characteristic ways of attacking PRAS; and a rebuttal of the attack using a credible source argument appropriate to the topic concerned. Further discussion of credible source as an argument scheme can be found in [28].

4   Case Study: Speed Cameras

The formal machinery using the AATS+V is intended to provide the basis for the specification of semantic models which enable arguments about policy proposals to take place. We now consider the general process of policy making, and show how policies can be modelled and argued about, using a running example. The example is an issue in UK Road Traffic policy, previously used as an e-participation example in [3,5,10]. The number of fatal road accidents is an obvious cause for concern, and in the UK there are speed restrictions on various types of road, in the belief that excessive speed causes accidents. The policy issue which we will consider is how to reduce road deaths.

The starting point of policy making is when a policy issue on a particular topic is identified and the relevant governing body wishes to launch a consultation to solicit views on the issue. Since there is no specific commitment to a particular action at this stage, a *Green Paper* on the issue will be released publicly. The Green Paper is intended to encourage debate, with a view to interested parties, such as unions, pressure groups, think tanks, companies, universities etc., putting forth their views and comments on the issue, which they submit as formal responses. Considering our running example, the Green Paper would solicit opinions on the issue of what to do to reduce road deaths.

At this deliberative stage of the process, typically a wide range of proposals is put forward representing the different perspectives of different parties with different expertise, interests and values on the issue. For these to inform policy making, the relevant government department must analyse them to identify relevant facts, theories, interests and values, trying to synthesise them into some coherent form which can provide the basis of deliberation as to the policy to recommend in the subsequent *White Paper*. A White Paper sets out a concrete policy intended to form the basis of legislation and its justification. Again comments are sought from interested parties on the White Paper, but now with this rather specific focus. In short, when moving from the Green Paper to the White Paper, the government department tries to make sense of the alternative views submitted to try to produce a coherent picture of the domain of interest. Of course, this sense-making is not at present done using any formal apparatus. We argue, however, that such sense-making could be facilitated by formally representing the alternative views as AATS+V models, then reasoning with these models using argumentation schemes. This would clarify the alternative positions on the policy, force reconciliation of any incompatible views, and provide an integrated summary of the consultation. This aspect is discussed in more detail in [3].

4.1   Constructing Semantic Models of Policies

To fully describe a model using the AATS+V we need to specify the various components of the structure. We need the set of propositions Φ with which we can identify the possible member states of Q. Since if there are n elements in Φ there may be 2^n elements in Q, it is desirable to keep Φ as small as possible and only include propositions if they are definitely relevant to the problem. Given Φ, we can constrain the size of Q by identifying logical relationships between members of Φ, such that for $p_1, p_2 \in \Phi$, $\neg(p_1 \wedge p_2)$, which will allow the elimination of certain states. We need to give the set of agents, Ag, the actions they can perform, and any values inherently promoted or demoted by the performance of the action. Again, in order to keep the number of joint actions within reasonable bounds, we will need to be as frugal as possible in including agents and actions: n agents, each with m actions, give rise to n^m potential joint actions. Again this is an upper bound: some pairs of actions may be incompatible and so give rise to no joint action. Finally, we need a transition matrix expressing ρ, τ and δ. This matrix comprises a row for each state in Q and a column for each joint action in J. An entry in a cell indicates that the preconditions for the joint action are satisfied, and comprises a triple consisting of the state reached if that joint action is executed, the set of values promoted, and the set of values demoted. These transitions are a representation of a causal theory which explains the effects of various actions, and an evaluative theory which tells us when values are promoted and demoted.[10]

Returning to our running example, we suppose that we are trying to develop a policy to reduce road traffic deaths and have received responses to a Green Paper from which we will extract the various components of the AATS+V. As there may be alternative responses, we may need to create alternative models, or use the responses to build incrementally a complex model which represents the sum of the policy deliberations. The representation process is described in detail in [3], and also discussed with reference to a different domain in [18].

For example, one response to the Green Paper issue put forward by those concerned about road safety might be that we install and operate speed cameras at strategic points. The speed cameras automatically photograph speeding cars, and the photographs are subsequently used to identify the car and issue speeding tickets to the drivers; we will use the installation of the cameras to refer to this overall process. There is evidence from other countries and pilot studies that this measure can be effective. So we might propose the following as the intended meaning of the response: *The government should install speed cameras to reduce road deaths, which will promote the value of Life.* However, we want to argue about policy using our practical reasoning argumentation scheme, which explicitly references circumstances and consequences. The response just given is elliptical, having only the action and the value. So to be compatible with PRAS, we need to add the current circumstances (that road deaths are too high, and that speeding is rife), and a consequence (that there will be fewer accidents and so fewer deaths). There is still some magic here, however: it is not the speed cameras themselves that reduce the accidents: the belief is that speed cameras will cause motorists to observe the speed limits, that observing speed limits will reduce accidents, and this will lead to fewer deaths, and so we need to include motorists and how they change their behaviour in response to the policy in our model.

From this initial conceptualisation of the problem, we present an initial model in the form of the following AATS:

- $Q = \{q_0, q_1, q_2\}$. Although we have two propositions (and so four potential states) we model the assumption that a reduction of speeding will reduce road deaths and so ignore the state $r, \neg s$.
- $Ag = \{G, M\}$, where G is the government and M is motorist[11];
- $Ac_G = \{G_1, G_0\}$, which are the actions the government does or does not perform, respectively. $Ac_M = \{M_1, M_0\}$, which are the actions the motorist

---

[10] In order to keep matters simple we chose to restrict goals to elements of Φ and conjunctions thereof for both our tools. The machinery to handle more complex goals is fully described in [1].

[11] Where *motorist* is an abstraction to use the 'collective' interpretation of 'motorist'.

does or does not perform. Here G_1 is *operate speed cameras*, and M_1 is *cut speed*. G_0 and M_0 are, respectively, that the government and the motorist do nothing.

– $\Phi = \{r, s, \neg r, \neg s\}$. where r represents road deaths being high and s represents there being excessive speeding. While we informally also have have a proposition a representing a high accident rate, we assume, to keep the number of states down, that a and r can be taken as equivalent, since accidents and deaths are correlated;

– $V = \{L\}$. Our one value is saving lives.

– δ is such that $\delta(q_x, q_y, L) = +$, if r holds in q_x and $\neg r$ holds in q_y; − if $\neg r$ holds in q_x and r holds in q_y; and $=$ otherwise.

– π is a function such that $\pi(q_0) = \{r, s\}$, $\pi(q_1) = \{\neg r, s\}$, and $\pi(q_2) = \{\neg r, \neg s\}$;

– J_{Ag}, the set of all joint actions, is $\{j_0, j_1, j_2\}$, where j_0 is $< G_0, M_0 >$, j_1 is $< G_1, M_0 >$, j_2 is $< G_1, M_1 >$. We have eliminated one logically possible joint action by assuming that Motorists do not cut their speed if the government does nothing.

The model also requires the functions ρ (for action pre-conditions) and τ (for system transitions). We can express these as in a transition matrix shown in Table 1: an entry in a cell indicates the pre-conditions for the joint action are satisfied; the first argument is the state reached if that joint action is executed, the second is the set of values promoted, and the third is the set of values demoted; where no value is promoted or demoted, we have "_"; *null* means the pre-conditions of one or more of the component actions cannot be satisfied, so that joint action is not possible in that state. This is true of j_0 in q_2 in our example: in the case where the speed cameras have succeeded in reducing speeds, it is assumed that the government will continue to operate them, so that only the joint actions containing G_1 are possible in q_2.

Table 1. Initial Transition Matrix

	j0	j1	j2
q0	$\langle q0,\_,\_\rangle$	$\langle q0,\_,\_\rangle$	$\langle q2,+L,\_\rangle$
q1	$\langle q1,\_,\_\rangle$	$\langle q1,\_,\_\rangle$	$\langle q2,\_,\_\rangle$
q2	null	$\langle q0,\_,-L\rangle$	$\langle q2,\_,\_\rangle$

A second response might be from a group of people who dispute that excessive speeding is a factor in deaths. In order to represent that the effect of actions can be indeterminate, we introduce a third agent N, which is usually termed *nature*, and distinguish two joint actions containing the indeterminate action, depending on whether nature cooperates, (here, meaning that a reduction in speed has the desired effect on deaths), or nature does nothing, (which here means that a reduction in speed does not have the desired effect).

The second response was intended as an objection to speed cameras. A third response might, however, provide a rebuttal to this objection by saying that even if compliance with speed limits did not have a significant effect on accidents, it would still be worthwhile, since it would mean that there was increased compliance with the law, and that this is a value in itself (C).

Next we may need to add some additional aspects, considering the cost of the proposal and an alternative proposal involving education. Speed cameras cost money, and there is only a limited budget available for improving road safety. We therefore need to consider monetary matters. This will relate to a value B, which is demoted if the budget is exceeded and promoted if there is a surplus. Assuming we do have money to spend, we can cost our plan and interpret the action of introducing cameras as being the introduction of such speed cameras as the budget will allow. Where cameras are installed according to budget the action is neutral with respect to B and so the transition will be neutral with respect to B. If, however, motorists fail to respond to the deterrent effect of the cameras, continue to speed, and pay the fines, then, because we can easily identify and prosecute the speeders, income from fines will be greater than expected and the expenditure will be recouped.

For an alternative action, suppose there is a submission by a group who believes that introducing speed cameras will not reduce road deaths, but is very much in favour of reducing these deaths. They may argue that some other action (G_2) is required to be effective. For example, if we were to educate drivers, so that they were better aware of the effects of speed, and better able to handle their vehicles at speed, then we would expect to reduce accidents, and hence deaths. Thus the government's education of drivers would, it is argued, lead to a reduction in deaths whether or not speeding decreased, since motorists who continue to speed are better able to control their cars. The only problem is that education is more expensive than cameras and does not give rise to any revenue stream, and so this proposal would be over budget, demoting B.

All this gives the final AATS+V shown diagrammatically in Fig. 1.

To keep the set of actions small, the action used to represent education can also be used to represent any other government actions which it is claimed will lead to a reduction in accidents but which will exceed the budget, such as deploying increased numbers of traffic police to catch speeders. Note, however, that we now need to distinguish between speeding and accidents, and so require the fourth state where speeding continues, but deaths decrease, reachable by the joint action educating motorists who continue to speed.

Finally, we will consider responses to the Green Paper that are representative of arguments from Civil Liberties pressure groups. They argue that speed cameras, by revealing the location and movements of citizens, represent an unacceptable intrusion of privacy. This requires a new proposition (p) to represent the existence of the speed cameras making an excessive intrusion on privacy. This will be accompanied by an additional value, F representing civil liberties. This requires an extension to the model: adding p splits every state reachable by introducing cameras into two to distinguish states where privacy is respected from those where it is not.

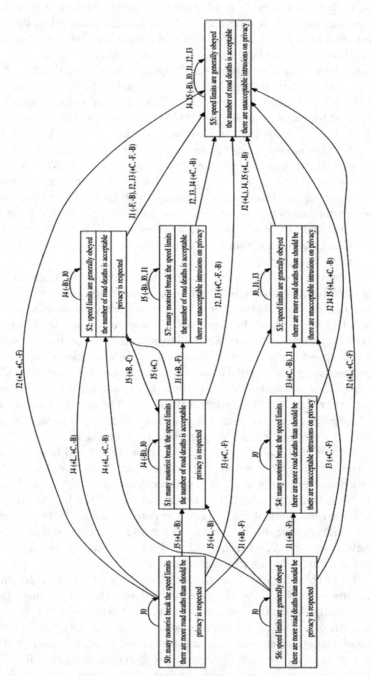

Fig. 1. AATS+V for Speed Camera Domain

Formally the components of this AATS+V are:

- $Q = \{q_0, q_1, q_2, q_3, q_4, q_5, q_6\}$.
- $Ag = \{G, M, N\}$, where G is the government, M is motorist and N is Nature.
- $Ac_G = \{G_2, G_1, G_0\}$, which are the actions of the government, respectively educate motorist, introduce cameras, and take no action. As before, $Ac_M = \{M_1, M_0\}$, and AC_N is $\{N_1, N_0\}$, depending on whether or not reducing speed also reduces deaths.
- $\Phi = \{r, s, p, \neg r, \neg s, \neg p\}$, where r represents road deaths being high, s represents there being excessive speeding and p represent unacceptable intrusions on privacy.
- $V = \{L, C, B, F\}$, as explained above.
- δ is such that $\delta(q_x, q_y, L) = +$, if r holds in q_x and $\neg r$ holds in q_y; $-$ if $\neg r$ holds in q_x and r holds in q_y; and = otherwise. $\delta(q_x, q_y, C) = +$, if s holds in q_x and $\neg s$ holds in q_y; $-$ if $negs$ holds in q_x and s holds in q_y; and = otherwise. $\delta(q_x, q_y, B) = +$, if the transition between q_x and q_y contains both G_1 and M_0; $-$ if the transition between q_x and q_y contains G_2 and = otherwise. $\delta(q_x, q_y, P) = +$, if $\neg p$ holds in q_x and p holds in q_y; $-$ if p holds in q_x and $\neg p$ holds in q_y; and = otherwise.
- π is a function such that states are interpreted as shown in Fig. 1.
- J_{Ag}, the set of all joint actions, is $\{j_0, j_1, j_2, j_3, j_4, j_5\}$, where j_0 is $< G_0, M_0, N_0 >$; j_1 is $< G_1, M_0, N_0 >$; j_2 is $< G_1, M_1, N_0 >$, j_3 is $< G_1, M_1, N_1 >$, j_4 is $< G_2, M_1, N_1 >$ and j_5 is $< G_2, M_0, N_1 >$.

The functions τ and ρ are shown in the transition matrix in Table 2.

Table 2. Final Transition matrix.

	j0	j1	j2
q0	⟨q0,-,-⟩	⟨q0,+B,-F⟩	⟨q5,+L+C,-F⟩
	j3	**j4**	**j5**
q0	⟨q6,+C,-F⟩	⟨q2,+L+C,-B⟩	⟨q3,+L,-B⟩

When the response period for the Green Paper closes, the opinion gathering ends and the policy analyst can then focus on the proposal to be chosen as the preferred option to be set out in the White Paper, forming the next part of the process. It is at this point that we envisage the Structured Consultation Tool described below being deployed.

4.2 Implementation

Once we have identified the elements of an AATS+V, implementation is straightforward. The AATS+V is described by representing three relations: states, joint actions, and transitions. These relations are represented in data structures appropriate to the language of choice. In Prolog they would be clauses: if using a database they would be tables. The Prolog prototype was described in [25] and the database version is described in [23] and, more fully, in [27]. Sample Prolog clauses for the above AATS+V would be:

```
state(0,1,3,6).
jointAction(j0,[do,nothing],[do,nothing],[have,no,effect]).
transition(1,0,5,j2,[1,c],[f]).
```

where state has an id, and a literal for each of r, s, and p; joint action has an id, and action for each of the three agents; and transition has an id, a source state, a target state, a set of values promoted and a set of values demoted. Additional relations can be used to provide additional information such as textual descriptions:

```
value(4,b,budget).
action(government,3,[educate,motorists]).
literal(1,1,[there,is,excessive,speeding],[]).
```

Now we can run appropriate queries to instantiate PRAS and attacks upon its instantiations. For example we can instantiate PRAS using the Prolog query:

```
argumentPro(A,S,R,V):-transition(ID,S,R,J,X,_),
                member(V,X),
                jointAction(J,A,_,_)
                ([government,should,A,in,S,to,reach,R,
                and,promote,V]).
```

and identify an attack based on the demotion of a value with:

```
argumentCon(A,S,R,V):-transition(ID,S,R,J,_,X),
                member(V,X),
                jointAction(J,A,_,_),
          write([government,should,not,A,in,S,to,avoid,R,
                              which,would,demote,V]).
```

It is a simple matter to write equivalent queries in SQL if using a database (see [27]).

5   Justifying and Critiquing Policies

In this section we will describe our applications built using the apparatus described above. We have two tools: one, the SCT, presents a justification of a policy and receives feedback on which points the citizens agree with and which points they disagree with. The second tool, the Critique Tool (CT), reverses the roles; this tool solicits a proposal from the citizen and then provides a critique from the government perspective.

The idea is that internally the system will operate by instantiating PRAS and attacks upon these instantiations based on the AATS+V, which represents the domain model. All of this will, however, be hidden from the users who will be presented with a series of screens presenting these justifications and attacks, and users will be asked to answer "yes" or "no" to indicate their agreement or disagreement. Once given, these answers can be interpreted in terms of the model, so that the statements agreed with become justifications and attacks on justifications. This further allows the responses to be aggregated so that it becomes clear what the specific strengths and weaknesses (as perceived by the citizens) of a policy justification are.

Both tools are implemented in MySQL and then embedded in PHP to provide access over the internet. The tools are (May 2014) available at

- http://impact.uid.com:8080/impact/ and
- http://cgi.csc.liv.ac.uk/~maya/ACT/

5.1   Structured Consultation Tool (SCT)

The role of the SCT is:

- to present the justification of a policy to members of the public;
- to allow members of the public to disagree with certain specific points of that justification;
- to present Credible Source arguments to justify the points disagreed with.

In this way the popularity of the policy can be gauged, and, if it is not supported, the reasons why it is not popular identified. The AATS+V can be used to instantiate a justification for the policy for presentation to the public using the SCT. Feedback on the argument and the model used is then sought, concerning disagreements and omissions, the assumptions made, and the ordering of values chosen.

After an initial statement of the justification, participants who disagree are led through a series of screens to identify the particular points at which they disagree, or want further justification. Further justification of specific points is given by a "digression" which presents (and receives feedback on) an appropriate credible source argument.

- *Screen 1* asks about the current state. For each proposition in the current state, the participant is invited to agree or disagree that it is the case. This corresponds to an epistemic challenge on the beliefs as to what is current the case. If there is disagreement, evidence is presented (e.g. accident statistics). If the participant remains unconvinced, the argument supporting the premise can be critiqued. The first screen also asks the participant to list any other relevant facts that need to be considered. To give an example of the look and feel of the system, a screen shot is shown in Fig. 2[12].

- *Screen 2* asks questions such as "Do you agree that reducing road deaths promotes life?", so that each of the labellings of the relevant transitions can be questioned. This effectively challenges the way the δ function has been defined in the AATS+V.
- *Screen 3* relates to the states reached by a transition. Participants are asked if the propositions claimed to be true in the next state will indeed result from the action. This challenges the underlying causal model relating actions and outcomes in the AATS+V. Disagreement will result in an argument justifying that transition being shown, and either participants will accept this and return, or be led through a critique of this further argument justifying the causal relationship. This screen also offers the opportunity to identify unstated consequences of the action thought relevant and undesirable.
- *Screen 4* offers a range of other actions (such as G_2 in the speed camera example) which participants may think achieve the aims of the policy. Selecting one of these leads to the reason for rejecting it (in the example, that this action would be beyond the available budget). Any other alternative actions not included in the AATS+V supported by participants may be entered as free text.
- *Screen 5* asks about values: whether participants endorse the values used, or want other values considered, and gives the opportunity to express their ordering of values. This is to explore whether it is the desirability of the policy rather than its effectiveness and feasibility that is being challenged. Such challenges are intrinsically subjective, whereas the earlier challenges can be seen as objective.

When participants have submitted their opinions, we can see whether our proposed policy commands popular support and, if not, exactly why not. Screen 1 should confirm that the number of deaths and accidents are seen as a problem, and asks for any factors other than speeding which may be seen as a cause of the problem. A substantial write-in for poor lighting, coupled with later comments, would indicate that a different approach has popular support. Screen 2 is about the link between goals and values. It may be that people disagree that cameras represent an unacceptable intrusion on privacy, which would be good news for advocates of the policy of introducing cameras. Screen 3 allows the underlying causal model to be questioned. This is the opportunity to deny that speeding causes accidents, and, for example, to offer poor lighting as an alternative cause. Screen 4 gauges support for G_2, and is where people may suggest other alternatives, such as improved lighting. The acceptability of the budgetary argument against G_2 is also indicated by the reception of the argument against G_2 in the digression. Finally Screen 5 tests our assessment of value priorities. We ranked life above privacy; this may be endorsed or disputed. The advantages of the SCT over current tools are:

---

[12] The application shown in the screenshot is that addressed by the IMPACT project, concerning a copyright topic. See [18].

- Justification is structured;
- Both citizens and officials are helped to make good arguments;
- Interaction has a natural flow;
- Replies are cogent and to the point;
- Specific points of disagreement are identified;
- No training or theory is required;
- Users only have to answer yes or no;
- The structure allows for replies to be related and aggregated;
- Which aspects of the policy require change or better explanation are identified.

5.2 Critique Tool

The second tool reverses the roles of the SCT: now it is the citizen who is providing the proposal and the justification and the tool which supplies a critique by finding objections using the model.

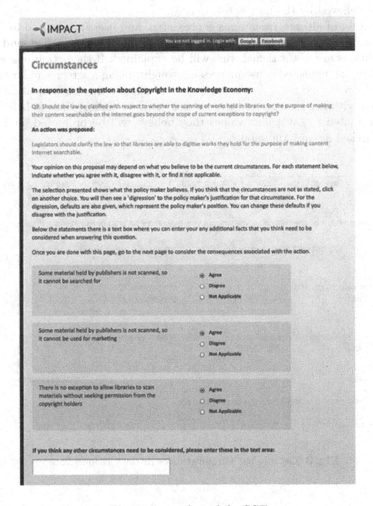

Fig. 2. Screenshot of the SCT

The first thing to do is to get the proposal. This is done by guiding users step by step through the instantiation of PRAS with respect to their conceptualisation of the domain, but requiring only "yes" or "no" responses. Thus users are presented with the screen shown as Fig. 3.

Note that the relevance of circumstances is determined by what the government considers relevant. This can, of course, be seen as a weakness, but it reflects that proposing a policy will only be persuasive if it is couched in terms acceptable to its audience [6], which, in the context of this tool, is the government rather than the citizen. The user having indicated which propositions are believed, the responses will be checked against q_0. The answers may be agreed, as in Fig. 3, or arguments justifying the different beliefs presented. This could be done using credible source arguments, as the digressions of the SCT, or simply by presenting some justifying text or web resource, as in the prototype critique tool. Users may at this point change their minds or stick with their original beliefs, in which case the consultation will proceed using their assumptions (i.e. the state believed by the user is taken as q_0.)

Next a set of alternative actions is presented, and users are invited to choose an action. The action is checked for its pre-conditions being satisfied, and if the pre-conditions are not satisfied, this will be explained. If the action is accepted as possible, the expected consequences are sought, using a screen similar to that relating to circumstances. Again any points of disagreement are identified, and supported with arguments, and the users invited to change their beliefs. Users are then invited to say which values they believe will be promoted, and again this is checked against the model.

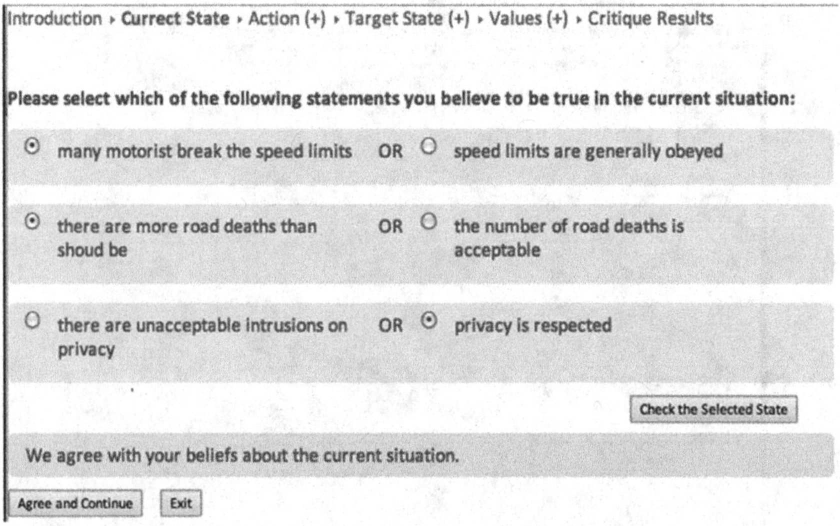

Introduction ▸ **Currect State** ▸ Action (+) ▸ Target State (+) ▸ Values (+) ▸ Critique Results

Please select which of the following statements you believe to be true in the current situation:

⊙ many motorist break the speed limits   OR ○ speed limits are generally obeyed

⊙ there are more road deaths than   OR ○ the number of road deaths is
shoud be   acceptable

○ there are unacceptable intrusions on   OR ⊙ privacy is respected
privacy

Check the Selected State

We agree with your beliefs about the current situation.

Agree and Continue   Exit

Fig. 3. Getting the Circumstances in the Critique Tool

This completes the solicitation of an instantiation of PRAS, and so at this stage we will have an argument justifying a policy which is valid according to the model. Although this means that the argument is a valid argument, there may be a number of reasons why it might be considered unacceptable:

- It may have undesirable side effects, demoting values;
- There may be other, perhaps better, ways of promoting the values;
- It may be possible to promote different, perhaps preferable, values;
- Other agents may not respond as anticipated.

All of these objections can be identified by posing simple queries to the model, and are then presented to the user using the screen shown in Fig. 4.

The strengths of the Critique Tool are that the argumentation scheme, critical questions and the underlying model together allow a systematic and intelligent critique of a proposal to be automatically generated from the model. This:

- Challenges assumptions and factual errors;
- Provides supporting arguments if required;
- Offers alternative ways of promoting the desired values, and alternative values that can be pursued in the circumstances, and identifies flaws in the arguments, and any potentially damaging side effects and risks posed by others not behaving as anticipated.

Like the SCT, the responses made using the Critique Tool can be interpreted, stored and aggregated, in terms of the argumentation scheme and the AATS+V.

5.3 Linking to Other Sites

As well as their primary purpose of getting feedback from citizens on policy issues, the tools can also be a means for members of the public to explore and learn about the various issues. The fact that the tools are embedded in the internet means that there is ready access to a wealth of information. The two tools offer different ways of justifying claims: the SCT using digressions to present arguments based on credible sources which can be interacted with, and the Critique Tool referring on to the credible sources themselves. It would also be possible to combine these methods; first presenting the credible source as a web page and then summarising it as an argument that can be critiqued.

A further possibility is to present information resources putting the pros and cons of the various points to the user *before* they are asked to express their beliefs and opinions. This puts the user in the position of an arbiter rather than a proponent of a particular side of the debate. Such a tool might be especially useful at the Green Paper stage of a policy consultation.

5.4 Evaluating the Responses

Both tools serve to collect responses which can be organised as arguments and counter arguments. This means that as well as purely numeric processing, which

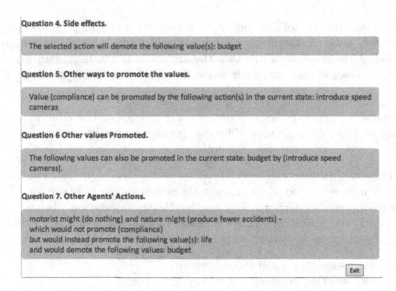

Fig. 4. Possible Objections to the User's Proposal

would enable us to say, for example, that 90% of respondents agree that there are too many road deaths, and that 87% agree that introducing speeds cameras would ameliorate this, we can take advantage of the argument structure. In computational argumention, since the introduction of argumentation frameworks by Dung [11], much has been done to determine the status of arguments in a framework of conflicting arguments and counter arguments. These techniques have also been applied to arguments which vary in strength according to the preferences of their audiences, in particular preferences based on value orderings (Value-based Argumentation Frameworks) [7] and preferences which can themselves be argued for (Extended Argumentation Frameworks) [17]. The idea is to find consistent sets of arguments which support one another and so form coherent positions on an issue, or identify which arguments require particular assumptions. These techniques can support both the initial policy choice, and the analysis of responses received from our tools. This argumentation oriented assessment of responses is explored in [9] and formulation and assessment of arguments about speed cameras using value-based argumentation frameworks is described and discussed in [3].

6   Concluding Remarks

A major problem with current e-participation systems is organising the replies for comparison, aggregation and assimilation. One answer to this is to make use of a well defined argumentation structure to organise policy justifications and critiques of these justifications. We have described:

- An argumentation scheme to structure justification and critiques;
- A semantical structure for models to underpin this scheme;
- A tool to facilitate a precise critique of the scheme;
- A tool to elicit a well formed justification and generate an automatic critique.

Both tools operate on the same underlying models of argumentation and of the domain. Some small scale evaluation exercises have been conducted with the SCT, and its earlier incarnation as described in [9]. Initial feedback about early versions of the SCT was positive about the aims of the system, in terms of supporting participatory democracy, and the ease with which the tool could be learned and used. However, users also expressed a desire to be able to put forward their own proposals. This identified the need for the Critique Tool that was subsequently developed.

Both tools are currently research prototypes and require evaluation in a serious situation concerning a genuine, live, policy issue. Building the model does require a considerable investment to time and expertise, but this is true of the conventional consultation process as well. We would argue that the potential gains from using such tools in terms of the quality of the feedback received, and the ease of analysing the feedback, would justify the effort required when undertaken as part of a consultation process. The purpose of the prototypes is to demonstrate the potential of using a well structured form of argumentation to present positions and receive feedback: we would anticipate that evaluation would identify opportunities for the tools to refined and extended.

Acknowledgements. This paper represents a consolidated version of work carried out at the University of Liverpool on the European project IMPACT (FP7-ICT-2009-4 Programme, Grant Number 247228). The views are those of the authors. It is a revised and much extended version of a keynote talk given by the first author at DEXA 2013 in Prague [8]. It draws on a series of earlier papers: especially [3,25,26,28]. We would particularly like to thank our colleagues Maya Wardeh, who did much of the implementation, Dan Cartwright, who explored an earlier version of the Structured Consultation Tool (Parmenides) in his PhD thesis [9], and colleagues on the IMPACT project. The work described here has its ultimate origins in [13], also presented in Prague.

References

1. Atkinson, K., Bench-Capon, T.: States, goals and values: Revisiting practical reasoning. In: Proceedings of Argmas 2014 (2015, In Press)
2. Atkinson, K., Bench-Capon, T.J.M.: Practical reasoning as presumptive argumentation using action based alternating transition systems. Artif. Intell. **171**(10–15), 855–874 (2007)
3. Atkinson, K., Bench-Capon, T.J.M., Cartwright, D., Wyner, A.Z.: Semantic models for policy deliberation. In: Ashley, K.D., van Engers, T.M. (eds.) Proceedings of the Thirteenth International Conference on Artificial Intelligence and Law, ICAIL, pp. 81–90. ACM, Pittsburgh (2011)
4. Atkinson, K., Bench-Capon, T.J.M., McBurney, P.: Computational representation of practical argument. Synthese **152**(2), 157–206 (2006)

5. Bench-Capon, T., Prakken, H.: A lightweight formal model of two-phase democratic deliberation. In: Proceedings of JURIX 2010, pp. 27–36. IOS Press (2010)
6. Bench-Capon, T.J.M.: Agreeing to differ: modelling persuasive dialogue between parties with different values. Informal Log. **22**, 231–246 (2002)
7. Bench-Capon, T.J.M.: Persuasion in practical argument using value-based argumentation frameworks. J. Log. Comput. **13**(3), 429–448 (2003)
8. Bench-Capon, T.: Structuring E-participation in policy making through argumentation. In: Decker, H., Lhotská, L., Link, S., Basl, J., Tjoa, A.M. (eds.) DEXA 2013, Part I. LNCS, vol. 8055, pp. 4–6. Springer, Heidelberg (2013)
9. Cartwright, D.: Digital decision-making: using computational argumentation to support democratic processes. Ph.D. thesis, University of Liverpool (2011)
10. Cartwright, D., Atkinson, K.: Using computational argumentation to support e-participation. IEEE Intell. Syst. **24**(5), 42–52 (2009)
11. Dung, P.M.: On the acceptability of arguments and its fundamental role in nonmonotonic reasoning, logic programming, and n-person games. Artif. Intell. **77**, 321–357 (1995)
12. Gordon, T.F., Karacapilidis, N.I.: The zeno argumentation framework. In: Sixth International Conference on Artificial Intelligence and Law, pp. 10–18 (1997)
13. Greenwood, K., Bench-Capon, T.J.M., McBurney, P.: Structuring dialogue between the people and their representatives. In: Traunmüller, R. (ed.) EGOV 2003. LNCS, vol. 2739, pp. 55–62. Springer, Heidelberg (2003)
14. Karacapilidis, N.I., Papadias, D.: Computer supported argumentation and collaborative decision making: the hermes system. Inf. Syst. **26**(4), 259–277 (2001)
15. Kunz, W., Rittel, H.W.J.: Information science: On the structure of its problems. Inf. Storage Retrieval **8**(2), 95–98 (1972)
16. Macintosh, A., Gordon, T., Renton, A.: Providing argument support for eparticipation. J. Inf. Technol. Polit. **6**(1), 43–59 (2009)
17. Modgil, S.: Reasoning about preferences in argumentation frameworks. Artif. Intell. **173**(9–10), 901–934 (2009)
18. Pulfrey-Taylor, S., Henthorn, E., Atkinson, K., Wyner, A., Bench-Capon, T.J.M.: Populating an online consultation tool. Leg. Knowl. Inf. Syst. JURIX 2011, 150–154 (2011)
19. Reed, C., Rowe, G.: Araucaria: Software for argument analysis, diagramming and representation. Int. J. Artif. Intell. Tools **13**(4), 983 (2004)
20. Schneider, J., Groza, T., Passant, A.: A review of argumentation for the social semantic web. Semant. Web **4**(2), 159–218 (2013)
21. Searle, J.R.: Rationality in Action John R. Searle A Bradford Book. MIT Press, London (2001). Please check the edit made in reference [21]
22. Walton, D.: Argumentation Schemes for Presumptive Reasoning. Lawrence Erlbaum Associates, Mahwah (1996)
23. Wardeh, M., Wyner, A., Atkinson, K., Bench-Capon, T.J.M.: Argumentation based tools for policy-making. In: The 14th International Conference on Artificial Intelligence and Law, pp. 249–250. ACM Press (2013)
24. Wooldridge, M., van der Hoek, W.: On obligations and normative ability: towards a logical analysis of the social contract. J. Appl. Log. **3**(3–4), 396–420 (2005)
25. Wyner, A., Atkinson, K., Bench-Capon, T.J.M.: Critiquing justifications for action using a semantic model: Demonstration. In: Computational Models of Argument - Proceedings of COMMA 2012, pp. 503–504. IOS Press, (2012)

26. Wyner, A., Atkinson, K., Bench-Capon, T.J.M.: Opinion gathering using a multi-agent systems approach to policy selection. In: van der Hoek, W., Padgham, L., Conitzer, V., Winikoff, M. (eds.) Proceedings of the 11th International Conference on Autonomous Agents and Multiagent Systems, AAMAS, pp. 1171–1172. IFAAMAS, Valencia (2012). Please check the publisher location for reference [26]

27. Wyner, A., Wardeh, M., Bench-Capon, T.J.M., Atkinson, K.: A model-based critique tool for policy deliberation. In: The Twenty-Fifth Annual Conference on Legal Knowledge and Information Systems - JURIX 2012, pp. 167–176. IOS Press (2012)

28. Wyner, A., Atkinson, K., Bench-Capon, T.: Towards a structured online consultation tool. In: Tambouris, E., Macintosh, A., de Bruijn, H. (eds.) ePart 2011. LNCS, vol. 6847, pp. 286–297. Springer, Heidelberg (2011)

Horizontal Business Process Model Integration

Klaus-Dieter Schewe[1,2](✉), Verena Geist[1], Christa Illibauer[1], Felix Kossak[1], Christine Natschläger-Carpella[1], Theodorich Kopetzky[1], Jan Kubovy[2], Bernhard Freudenthaler[1], and Thomas Ziebermayr[1]

[1] Software Competence Center Hagenberg, Hagenberg im Mühlkreis, Austria
{kd.schewe,verena.geist,christa.illibauer,felix.kossak,
christine.natschlaeger,theodorich.kopetzky,bernhard.freudenthaler,
thomas.ziebermayr}@scch.at
[2] Johannes-Kepler-University Linz, Linz, Austria
{kd.schewe,jkubovy}@faw.jku.at

Abstract. Modelling business processes in general is a complex endeavour, as many different aspects such as the control flow, the management of data, event and message handling, actors and interaction, exception handling, etc. have to be taken into account, all of which require different models. This paper focuses on the horizontal integration of models for control flow, message flow, event handling, interaction, actors, data and exception handling. The method is based on Abstract State Machines (ASMs), which are used to formally define the semantics of each of the individual models. Throughout the process rigorous quality assurance methods will be applied.

1 Introduction

Modelling information systems in general is a complex endeavour, as systems comprise many different aspects such as the data, functionality, interaction, distribution, context, etc., which all require different models. In addition, models are usually built on different levels of abstraction and the switch from one of these levels to another one may cause mismatches. Horizontal model integration refers to the creation of system models by successive enlargement, whereas vertical model integration refers to the systematic, seamless refinement process of high-level abstract (conceptual) models down to running systems.

In this paper we concentrate on the horizontal integration of business process models following the approach sketched in [27]. Taking a meromorphic view we consider complex systems as aggregations of parts, each requiring a different model. Then the extension of one submodel by another one is formally handled by refinement capturing interfaces and overlaps in a consistent way. Key questions to be addressed concern the provision of a clear semantics for the integration, the

The research reported in this paper was supported by the European Fund for Regional Development as well as the State of Upper Austria for the project *Vertical Model Integration* within the program "Regionale Wettbewerbsfähigkeit Oberösterreich 2007–2013".

© Springer-Verlag Berlin Heidelberg 2015
A. Hameurlain et al. (Eds.): TLDKS XVIII, LNCS 8980, pp. 30–52, 2015.
DOI: 10.1007/978-3-662-46485-4_2

understanding of the information capacity of integrated models through notions of dominance and equivalence, and the integration into the process of requirements elicitation, refinement, validation and verification.

Business process modelling has a long tradition in research with many modelling approaches such as BPMN [30], YAWL [29], ARIS [26] or S-BPM [11], just to mention a few[1]. Syntactical errors can be checked with the help of an ontology defining the concepts in BPMN [18]. However, a key concern is that though all methods claim to have reached a high level of maturity, semantics [1,10,32], flexibility [25] and adequacy for the problem domain [31,33] are still matters of concern. Surprisingly, still many relevant aspects of business processes are not well covered, e.g. data handling or exeption handling, while others are overloaded. In other words, the issue of semantics is still open as discussed in detail by Börger in [4]. Our own research started from BPMN and so far closed several semantic gaps, which will be reported in the monograph [16].

1.1   Our Approach

With respect to horizontal model integration it is common to start with the *control flow model*, i.e. a business process is described in an abstract way by a set of activities and gateways, the latter ones for splitting and synchronisation, plus start and termination events. Depending on whether one, all or an arbitrary selection of (outgoing) paths are enabled in splitting gateways, we adopt the common distinction between XOR-, AND- and OR-gateways with an analogous distinction for the synchronisation gateways. However, this terminology is in a sense misleading, as there need not be a well-nested structure, in which a splitting-gateway corresponds to exactly one synchronisation gateway. This is one of the reasons, why we formalise the semantics of each of the constructs by means of Abstract State Machines (ASMs, [7]). As a state-based rigorous method, ASMs support the unambiguous capture of the semantics [5,8], in particular for OR-synchronisation [6]. Furthermore, on grounds of ASMs necessary subtle distinctions and extensions to the control flow model such as counters, priorities, freezing, etc. can be easily integrated in a smooth way. All constructs found in a control flow model are supposed to be executed in parallel for all process instances.

The control flow model is then extended by a *message model* and an *event model*. For this refinement in ASMs – mainly conservative extensions – are exploited [15]. In particular, the ground specification of firing conditions that depend on the state of the control flow, data, events and resources and actions that update this state [9] requires that only conditions and actions are refined. While messages are easily captured by means of specifications of sender and receiver, it becomes more subtle to define details such as synchronised vs. asynchronised messaging, delivery failure, rejection, message box overflows, etc. In our approach the ASM-based specification of messaging from S-BPM [11] has been adopted. For the event model it is necessary and sufficient to specify

---

[1] The survey in [24] tries to give a comparative evaluation.

what kind of events are to be observed, which can be captured on the grounds of monitored locations in ASMs, and which event conditions are to be integrated into the model.

The next horizontal extensions concern the *actor model*, i.e. the specification of responsibilities for the execution of activities (roles), as well as rules governing rights and obligations. This leads to the integration of deontic constraints [17,22], some of which can be exploited to simplify the control flow [20,21] or to handle optionality [19]. In this way subtle distinctions regarding decision-making responsibilities in BPM can be captured. Horizontal model integration through refinement is then extended towards an *interaction model* and a *data model*. For this, an abstract dialogue model is adopted (see [28] or similarly [12,14]) capturing interaction by means of operations on views that are defined on top of a database schema. In this way the data model results from view integration, but global consistency has to be addressed, as a global database infers dependencies between activities that are not visible on the control flow level.

Finally, an *exception handling model* has to be integrated to complete the horizontal integration picture. This is still in a preliminary state in our work. Overall, the general idea is that an exception is a disruptive event that requires partial rollback and depending on the state the continuation with a different subprocess.

1.2   Outline of This Article

In Sect. 2 we discuss simple control flow specifications. While syntactically we stay close to the BPMN approach with respect to activities and gateways we define semantics on grounds of ASMs and discuss some fundamental problems. We also show simple verification examples. In Sect. 3 we continue this discussion of control flow focussing now on extensions by flags and counters and their use. We also extend the discussion on verification. Then Sect. 4 is dedicated to messages and events, by means of which the control flow will be enriched by additional conditions and actions. This is followed by a discussion of data handling in Sect. 5, where we stress the importance of a two-layered approach separating the external data handling by means of dialogues and views supporting the activities in the control flow and the internal data handling by means of underlying databases, the associated problems of tansaction handling and view updates, and the impact on the semantics of the business processes. This is taken further in Sect. 6 addressing actors, roles and associated deontic concepts of permission and obligation. In this context we also briefly discuss principles of exception handling. We conclude in Sect. 7 with a brief summary and a discussion of further extensions needed with respect to horizontal business model integration. We also emphasise the role of complementary vertical business process model integration, which is outside the scope of this article.

2   Simple Control Flows

In this section we discuss the semantics and verification of simple control flow specifications as they appear in most control-oriented BPM specification methods, e.g. in BPMN. Our presentation stays close to BPMN, but we question some semantic declarations and require concretisations.

2.1   Components of Simple Control Flows

Basically, a **control flow specification** is a directed graph specifying the sequencing of (basic, i.e. atomic) activities. To be able to illustrate our very fundamental approach to semantics, let us for the moment restrict ourselves to only the following core components that are permitted in a simple control flow:

- **Activities**: An activity marks that someone has to execute a particular task such as compiling an order, writing a review, endorsing an application, preparing a delivery, composing a bill, etc. We consider activities to be atomic, but nonetheless the execution may need some time, and for the time being the duration of the execution is not specified. We only assume that eventually an activity that has been activated will also be terminated.
- **Start** and **End events**: A start event simply marks that a process will be initiated here. There is only one edge going out of a start event. A start event marks the termination of the process. There may be more than one edge leading to an end event.
- **Gateways**: A gateway enables the specification that the control flow is *split* into several branches, or analogously several branches are *synchronised* in a single continuing flow.

The simplest split gateways are *exclusive* and *parallel splits*. In the former case exactly one of the continuation paths will be followed, in the latter case all continuation paths will be followed. It is not foreseen to restrict gateways in a way that for each split there is a corresponding matching join. It may well be the case that a split is followed by other splits, where some branches will be synchronised, but it is also possible that no synchronisation is specified at all[2].

Analogously, there are *exclusive* and *parallel* synchronisation gateways, also denoted as *join* gateways. In the former case a single incoming flow will be passed on – we will have to discuss this later in this section. In the latter case all incoming flows must have been completed in order to continue with a single flow.

2.2   Examples

Let us look at some examples illustrated by Figs. 1, 2, 3 and 4. Though the core components in control flows are similar to those used by BPMN, we slightly

---

[2] Therefore, the notion of XOR-split and AND-split used as synonyms in the literature are misleading, as in general there is no well-defined bracket structure.

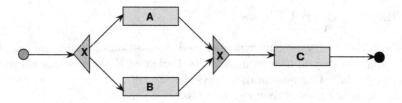

Fig. 1. Control flow with exclusive split and join

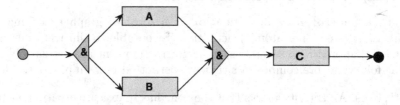

Fig. 2. Control flow with parallel split and join

modify the graph notation. In particular, we want to highlight the difference between split-gateways (one incoming edge and several outgoing ones) and synchronisation gateways (several incoming edges and only one outgoing one), so we use triangles instead of diamonds. In this way, we can reserve diamonds for gateways with several incoming and several outgoing flows, i.e. complex gateways. Furthermore, we use labels **X** and **&** for exclusive and parallel (split and join) gateways, respectively.

Example 1. Figure 1 shows a control flow with an exclusive split and a matching join. That is, after start the process will be either continued by activity **A** or activity **B**, followed by activity **C**. Similarly, Fig. 2 shows a control flow with a parallel split and a matching join. That is, after start the process will be continued by executing activities **A** and **B** in parallel, followed by activity **C**. So in both cases the informal semantics of the control flow is clear.

Example 2. This is not so clear for the control flow in Fig. 3, as the parallel split is matched by an exclusive join. Informally, this means that after start both activities **A** and **B** are executed in parallel, but what is the meaning of the synchronising exclusive join? If control is simply passed on as foreseen in BPMN, then activity **C** will be executed twice. If, however, this is not desired, the conrol flow could be considered to be incorrect, in which case this should be detected, and the control flow should not be permitted. Alternatively, it could still be considered to be correct, if the exclusive join gateway only passes the control on after the first completion of either **A** or **B**, whereas any follow-on enabling of the gateway would be ignored. In this case, however, the semantics of an exclusive join has to be specified in a way that permits to keep track, if still some information may arrive or not.

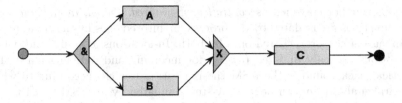

Fig. 3. Control flow with parallel split and exclusive join

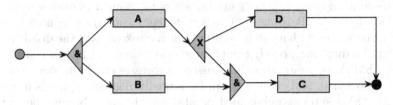

Fig. 4. Control flow without matching split and join gateways

Example 3. Finally, look at Fig. 4. In this case it seems informally clear that the control flow specification is not correct. After start activities **A** and **B** would be executed in parallel. After completion of **A** an exclusive split would either enable activity **D** or pass the control onto the parallel join gateway following activity **B**. That is, if **D** is enabled, this parallel join will wait forever, and activity **C** cannot be enabled. Only if control is not passed onto **D**, the parallel join would (after completion of **B**) pass control onto **C**.

Our examples already show two things: First, although the semantics of exlusive and parallel splits and joins seems to be informally clear, we have to be very precise about their meaning. In particular, it will be necessary to discuss exclusive joins. Second, the semantics must be defined in such a way that desired properties of the specified control flow can be verified. For instance, for the control flow in Example 3 we want to show that it may lead to a deadlock, i.e. the process may get stuck. In order to fulfil these requirements we will exploit Abstract State Machines (ASMs) for the rigorous definition of semantics

2.3 Abstract State Machines

ASMs are a rigorous, state-based method, where a specification can be considered as an iteration of a parallel execution of a set of rules of the form IF ⟨condition⟩ THEN ⟨action⟩. Conditions are evaluated on states, which are universal algebras, i.e. sets of functions (resulting from interpretations of function symbols). Actions initiated by the activities, gateways and the start and end events update these states, i.e. change the values of functions at certain locations (arguments). Functions of arity 0 capture the usual concepts of variables and constants in case these functions are declared to be dynamic (updateable) or static (non-updateable), respectively. Thus, functions can be *static*, *dynamic* or *derived*, the dynamic ones being further classified as *controlled, monitored, shared, in* and *out*. Controlled functions are updated by the process, monitored ones by the environment (e.g. in case of sensors), shared ones by both. In-locations are only read, where out-locations are only updated (as e.g. for incoming and outgoing mail). The theoretical background is the ASM thesis (Yuri Gurevich) according to which each parallel algorithm can be step-by-step simulated by an ASM [2, 3, 13].

In principle, any other equivalent formalism could be used, but ASMs have some advantages. In particular, we will see later that the concept of state is very important to easily capture features, where the semantics cannot be simply defined locally. If e.g. Petri nets [23] as a state-less formalism were used instead, this information would have to be captured in the tokens with the disadvantage of potential redundancy, overly complicated evaluation, and lack of clarity.

The ASM thesis guarantees that business processes can be modelled on any level of abstraction, and the levels can be formally related by means of refinement. As ASMs support unbounded parallelism, this may become helpful, as process instances run in parallel and also activities of each process instance do so. As states of ASMs are abstract, this can be exploited to easily capture more advanced concepts such as counting, priorities, coupling with databases, etc. Furthermore, the formal semantics forms a basis for rigorous methods for quality assurance by means of verification. Last, but not least ASMs have been applied in specifications and correctness proofs (even fully mechanised) for many application areas.

2.4   A Glimpse on ASM Semantics

Let us now sketch the ASM-based definition of semantics for simple control flows. We use a variable PROCESSES to capture at all time the (identifiers) of active process instances. Then the semantics of a start event is simply to create a new process instance:

LET $p = New$(process-id) IN PROCESSES := PROCESSES $\cup\{p\}$

Similarly, the semantics of an end event is to delete each incoming token – we will discuss the token model in the remainder of this section – and to remove a process instance from PROCESSES, if no more tokens are left:

IF *no_more_tokens*(p) THEN PROCESSES := PROCESSES $-\{p\}$ ENDIF

The basic rules for control flows can now be specified exploiting the parallelism and a token model. As already remarked, process instances run in parallel, which can be defined as follows:

FORALL p WITH $p \in$ PROCESSES DO run(p) ENDDO

Furthermore, also flow nodes, i.e. gateways, activities, etc. run in parallel:

FORALL f WITH $f \in$ FLOW-NODES(p) DO execute($f(p)$) ENDDO

For the specification of the control flow execution exploit a token model. For this we first capture the edges (aka sequence flows) in the underlying graph by means of functions *IncomingSequenceFlows* and *OutgoingSequenceFlows*, both defined on FLOW-NODES(p). Then we use an edge labelling function *tokensIn-SequenceFlow* associating a set of tokens with each sequence flow. Tokens can be modelled just by identifiers, but in addition we may define functions associating more detailed information with a token such as the associated process instance, the creating flow node, etc.

The semantics for flow nodes can then be specified in general as follows:

rule NodeTransition(flowNode) =
    IF
        controlCondition(flowNode) **AND** eventCondition(flowNode) **AND**
        dataCondition(flowNode) **AND** resourceCondition(flowNode)
    THEN
        parblock
            controlOperation(flowNode)
            eventOperation(flowNode)
            dataOperation(flowNode)
            resourceOperation(flowNode)
        endparblock

For the core model only the controlCondition and the controlOperation are relevant. This gives: execute($f(p)$) = NodeTransition($f(p)$).

Specialisation for Split Gateways. Let us now take a closer look at split gateways. For both exclusive and parallel split gateways the control condition is the same:

    splitControlCondition(flowNode) =
        Exists $e \in$ IncomingSequenceFlows(flowNode)
            With tokensInSequenceFlow(e) $\neq \emptyset$

Also, in both cases one incoming token has to be removed:

    removeIncomingToken(flowNode) =
        Choose $e \in$ IncomingSequenceFlows(flowNode)
            With tokensInSequenceFlow(e) $\neq \emptyset$
        Choose $t \in$ tokensInSequenceFlow(e)
            tokensInSequenceFlow(e) := tokensInSequenceFlow(e) $- \{t\}$

However, the production of outgoing tokens differs for exclusive and parallel split gateways. For exclusive split gateways just one outgoing token is produced

    produceOneOutgoingToken(flowNode) =
        Choose $e \in$ OutgoingSequenceFlows(flowNode)
        Let $t = $ **New**(token) **In**
            tokensInSequenceFlow(e) := tokensInSequenceFlow(e) $\cup \{t\}$

For exclusive split gateways one token is produced for every outgoing sequence flow:

    produceAllOutgoingToken(flowNode) =
        Forall $e \in$ OutgoingSequenceFlows(flowNode) **DO**

Let $t =$ **New**(token) **In**
   tokensInSequenceFlow(e) := tokensInSequenceFlow(e) $\cup \{t\}$
ENDDO

In summary, we obtain the following refinement for the exclusive split gateway:

exclusiveSplitTransition(flowNode) = NodeTransition(flowNode) **Where**
   controlCondition(flowNode) = splitControlCondition(flowNode) **AND**
   controlOperation(flowNode) =
      parblock
         removeIncomingToken(flowNode)
         produceOneOutgoingToken(flowNode)
      endparblock

Analogously, we obtain the following refinement for the parallel split gateway
parallelSplitTransition(flowNode) = NodeTransition(flowNode) **Where**
   controlCondition(flowNode) = splitControlCondition(flowNode) **AND**
   controlOperation(flowNode) =
      parblock
         removeIncomingToken(flowNode)
         produceAllOutgoingToken(flowNode)
      endparblock

Specialisation for Join Gateways. For exclusive join gateways the control condition is the same as for the split:

exclusiveJoinControlCondition(flowNode) =
   Exists $e \in$ IncomingSequenceFlows(flowNode)
   With tokensInSequenceFlow(e) $\neq \emptyset$

For parallel join gateways the control condition tokens must exist for all incoming sequence flows:

parallelJoinControlCondition(flowNode) =
   All $e \in$ IncomingSequenceFlows(flowNode)
   With tokensInSequenceFlow(e) $\neq \emptyset$

In both cases just one outgoing token is produced, for which the operation produceOneOutgoingToken(flowNode) can be reused.

For exclusive join gateways just one ingoing token is removed:

removeOneIncomingToken(flowNode) =
   Choose $e \in$ IncomingSequenceFlows(flowNode)
      With tokensInSequenceFlow(e) $\neq \emptyset$
   Choose $t \in$ tokensInSequenceFlow(e)
      tokensInSequenceFlow(e) := tokensInSequenceFlow(e) $- \{t\}$

For parallel join gateways one token is removed for every incoming sequence flow:

removeAllIncomingToken(flowNode) =
    Forall $e \in$ IncomingSequenceFlows(flowNode) **Do**
    Choose $t \in$ tokensInSequenceFlow(e)
        tokensInSequenceFlow(e) := tokensInSequenceFlow(e) $- \{t\}$
    Enddo

In summary, we obtain the following refinement for the exclusive join gateway:

exclusiveJoinTransition(flowNode) = NodeTransition(flowNode) **Where**
controlCondition(flowNode) = exclusiveJoinControlCondition(flowNode) **AND**
    controlOperation(flowNode) =
        parblock
            removeOneIncomingToken(flowNode)
            produceOneOutgoingToken(flowNode)
        endparblock

Analogously, we obtain the following refinement for the parallel join gateway:

parallelJoinTransition(flowNode) = NodeTransition(flowNode) **Where**
controlCondition(flowNode) = parallelJoinControlCondition(flowNode) **AND**
    controlOperation(flowNode) =
        parblock
            removeAllIncomingToken(flowNode)
            produceOneOutgoingToken(flowNode)
        endparblock

Let us finally look again at the semantics specified above for the exclusive join gateway. Roughly said, the specified semantics is "one incoming token will be removed and one outgoing token will be produced." Actually, this is equivalent to doing nothing: each token appearing on any incoming sequence flow is simply forwarded to the outgoing sequence flow – control flow is simply passed on. As we saw in Example 2 the effect may be that follow-on activities must be executed multiple times.

So the question is, whether this "empty" semantics is really the desired one? Alternatives to the specified semantics are the following ones:

– **True Join:** Only one incoming token is considered, all others are ignored, i.e. "the first one will be served". In this case other incoming tokens have to be deleted including those that still may arrive, but not via loops, so it is getting tricky, and additional information has to be kept in the state. We will discuss this alternative in the contexts of inclusive join gateways and flags in Sect. 3.
– **True XOR:** If it is possible to have more than one incoming token, this is considered an error. In this case it has to be verified that the erroneous situation may never occur.

2.5   A Glimpse on Verification

In our discussion of semantics we have now seen already several cases, where it is necessary to verify desired properties of control flow specifications. Examples of such properties are the following:

- **Liveliness:** Each started process will eventually terminate.
- **No deadlocks:** A flow flode may have to wait forever, though the process instance is not yet completed.
- **No redundancy:** Each activity or gateway can eventually be fired.
- **True XOR join:** Only one token may arrive at an incoming sequence flow of an exclusive gateway (third semantics).

Example 4. In Examples 1–3 we should see the following properties:

- Liveliness holds for the first three cases (empty semantics assumed) in Figs. 1, 2 and 3, only in the first two cases for the "true XOR" semantics, but not in the fourth case in Fig. 4.
- There are no deadlocks in the first three cases and no redundant flow nodes in any of the four cases.
- In the third case two tokens may arrive at the exclusive join, while in the fourth case it may occur that only one token arrives at the parallel join thus creating a deadlock.

In all these cases the proofs are rather obvious. Nonetheless, let us sketch examples of liveliness proofs.

First consider the control flow in Example 1 corresponding to Fig. 1. It can be easily seen that at any time there is at most one token, and that each path starting at the start node leads to the end node.

Next consider the slightly more complicated control flow in Example 1 corresponding to Fig. 2. In this case, if the parallel split is fired, then also the parallel join will fire. Then the proof can be reduced to the argumentation for the first example.

3   Control Flow Extensions

The discussion of different semantics for exclusive joins motivates to think about alternative gateways for synchronisation. Therefore, in this section we will introduce a few extensions to the simple control flow specifications considered so far. We will start with a discussion of inclusive split and join gateways. Informally, an *inclusion split* gateway will produce tokens on any subset of outgoing sequence flows, which can be easily specified by ASMs. Then an *inclusive join* must synchronise all incoming tokens that may arrive, but those that may arrive via a loop have to be excluded.

We adopt the semantics defined by Börger, Sörensen and Thalheim in [6], which provides a nice example for the use of flags in the state. Flags are de facto Boolean-valued locations, which can be used to define more complex control conditions and actions. As another extension we briefly sketch counters, i.e. integer-valued locations that can be used among others to capture priorities.

3.1 Inclusive Gateways

As stated above an inclusive join gateway has to synchronise all incoming tokens that may arrive (excluding loops). So, we need a different control condition. However, as the condition has to capture, if any token may still arive, the property is no longer local, i.e. it does not only depend on the token on incoming edges.

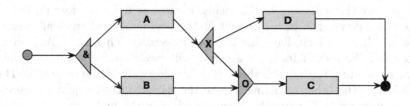

Fig. 5. Control flow with inclusive join gateway

Example 5. Let us modify the (erroneous) Example 3 illustrated in Fig. 4 by replacing the parallel join by an inclusive one. This is illustrated in Fig. 5. Intuitively, the informal semantics sketched above indicates that the control flow is now correct. As before after start the activities **A** and **B** will be executed in parallel and eventually completed. So the inclusive join will receive a token from **B** and wait, if another token arrives from the exclusive split following the completion of **A**. Now, if the exclusive split enables activity **D**, no such token may arrive at the inclusive join anymore, i.e. it will fire and thus enable activity **C** – so in this case both **C** and **D** will be executed. Otherwise, if the decision at the exclusive split is different, then **D** will not be executed, but the second token will appear at the other incoming edge of the inclusive gateway, which will fire and enable activity **C** – so in this case only **C** will be executed.

Example 6. Now consider a more complicated example as illustrated by the control flow in Fig. 6. After start the inclusive split gateway will select any subset of activities **A**, **B** and **C** to be executed in parallel. Regardless, which of these activities have been selected, they will eventually terminate. If **B** or **C** or both were selected, this will be synchronised again by the first inclusive join, and the

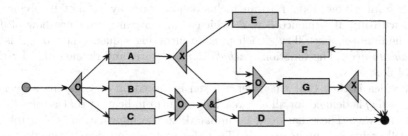

Fig. 6. Control flow with inclusive split and join gateways and feedback loop

follow-on parallel split will then pass tokens onto activity **D** and the second inclusive join. The latter one has two more incoming sequence flows. The one coming from activity **F** has to be ignored, as a token can only come this way, if it had passed already the (waiting) inclusive gateway, which is impossible. So, if **A** has been selected, executed, and the follow-on exclusive split does not select to pass on control to activity **E**, but to the second inclusive gateway, this one will fire.

To specify the semantics of the inclusive gateway it is therefore possible to proceed as follows. The inclusive split at the beginning "informs" all possible successors that are inclusive joins about its decision. That is, if **A** is in the selected set, the second inclusive gateway will receive a flag on the incoming flow corresponding to the path through **A**. Analogously, if one or both of **B** or **C** is selected, the other incoming flow of the second inclusive join will receive a flag in addition to flags on the incoming edges of the first inclusive gateway. If **A** is completed, but the following exclusive split passes control onto **E**, the corresponding flag can be removed, as no token may arrive via this path. Then an inclusive join can fire, if all tokens on incoming edges that could arrive actually have arrived. Then all tokens and flags will be removed and a token at the outgoing edge will be created.

Let us finally remark on the feedback loop. In case the second inclusive gateway fires, first **G** will be enabled. Then after completion, **F** might be enabled, in which case also a flag is created at the third incoming edge of the inclusive join. As no other flag could be set, this gateway could immediately fire again.

3.2   Semantics of Inclusive Gateways Using Flags

For inclusive splits the changes to exclusive or parallel gateways, respectively, are straightforward. Instead of creating exactly one token on one (or all, respectively) outgoing sequence flows, we select an arbitrary (in general non-empty) subset of these sequence flows and create tokens on all of them. In addition, all split gateways will have to create control flags for "reachable" inclusive join gateways, as we will discuss next.

So, for inclusive joins we want to exploit the paths from a split gateway of any type to a join gateway without any repetition of flow nodes. For this we can use a function *reachableJoin* defined for all outgoing sequence flows of split gateways assigning a set of sequence flows to them. More precisely, if there is a path from a flow node f to an inclusive join gateway f_o (not involving f_o itself) with initial sequence flow e (which is an outgoing sequence flow of f), then the final sequence flow e' (which is an incoming sequence flow of f_o) is in *reachableJoin*(e). The function *reachableJoin* is static and defined only by the control flow.

The idea is then that whenever a token is placed on an outgoing sequence flow e, a *flag* is defined for all $e' \in$ reachableJoin(e) indicating that a token may arrive this way. These flags can be updated, so the information, which tokens may still arrive is kept up to date. Thus, for each incoming sequence flow e of an inclusive join we define a function *flags* resulting in a set of flags, where a flag is defined as $t.o$, where t identifies a token, and o indicates the flow node, at which the token was generated. Then we have to refine all previous definitions of gateways:

- For each flow node the origin o of the generated tokens $t.o$ is an identifier for the gateway.
- All flags $t.o$ consumed by flow node f appearing in some $flags(e')$ with $e' \in$ $reachableJoin(e)$ and $e \in$ OutgoingSequenceFlows(f) will be replaced by a new flag $t'.f$, where t' is the new token produced on e.
- In particular, at the start s the generated token $t.s$ will be placed into all $flags(e')$ with $e' \in reachableJoin(s)$.

The refinement of the ASM specifications is straightforward.

Let us now look at the definition of the control operation and condition for inclusive join gateways. Informally, the control operation for an inclusive join is analogous to exclusive and parallel joins, as only one token is generated and the tokens needed in the control condition are removed. This includes the deletion of the flags. The control condition is simply to check that all tokens that may arrive have actually arrived:

inclusiveJoinControlCondition(flowNode) =
    All $e \in$ IncomingSequenceFlows(flowNode)
        With tokensInSequenceFlow$(e) = $ flags(e)

More details concerning the ASM specification for inclusive gateways can be found in [6]. It should be noted that in this particular case concerning the specification of inclusive joins the use of flags can be avoided as the inclusiveJoin-ControlCondition can be derived from the distribution of tokens, i.e. from the state. Furthermore, if multisets are used, identifiers for tokens may be preserved. However, it should also be noted that in any case the semantics of the inclusive gateway is no longer "local", as tokens in different parts of the specification have to be explored. This is simplified by the use of flags.

Nonetheless, flags associated with incoming (and outgoing) sequence flows can be used as a general extension mechanism. Another example for their use could be the definition of semantics of the "true join" semantics for exclusive joins "the first one coming is served".

3.3   Counters and Priorities

As we have already seen in our discussion of very simple examples (see Fig. 3), already with the (empty) semantics of exclusive joins it is possible to create multiple tokens associated with a single edge. Therefore, it appears natural to associate counters with incoming sequence flows to define complex control conditions requiring specified numbers of tokens on each incoming sequence flow to be consumed. Analogously, we may associate counters with outgoing sequence flows to define complex control operations generating specified numbers of tokens on each outgoing sequence flow. As such counters belong to the state, they may be subject of updates, e.g. updated by activities.

For instance, define how many reviews will be required by a certain application and how many positive ones will be needed for success. Then integrated split and join gateways as well as gateways with only one incoming and one outgoing sequence flow will be enabled.

This leads to several new verification problems. If complex control conditions involving counters are used, the question arises, if these conditions can always be satisfied. Another question is what happens with remaining tokens (as for exclusive joins). These tokens could be ignored or removed. Then again flags are needed for potentially more arriving tokens. Alternatively, left over tokens might be considered as a specification error. In case the soundness can only be checked at run-time, because counters can be updated, an exception may be raised.

Counters may also be used to count how often a flow node has fired for a particular process instance. This can be used to model different behaviour for the first, second, etc. run through the flow node. A special case for the use of such a run counter is the complex gateway in BPMN, for which the internal "state" can be modelled by the run counter, and control operations may affect the counter as well.

In a complex gateway with several incoming and several outgoing sequence flows priorities among several control conditions can be easily modelled. As the semantics of gateways (and other flow nodes) is defined by means of ASMs, any other complex conditions for splitting and synchronising the control flow can be modelled – if necessary, the state signature has to be extended.

4   Messages and Events

Messages and events are necessary extensions in business processes to fine-tune the specification. Basically, a *message* is defined by a *sender*, one or more *receiver(s)* and the *message content*. Sending a message is nothing more than a specific activity, but receiving a message can also be modelled as a specific activity, and the content of a message is created by the sender before sending – this is illustrated by the specific graphical notation for send and receive actions in Fig. 7. We distinguish a *signal* from a *message* by the main criterium that the signal is created continuously while sending. Messaging can be done in a *synchronous* or *asynchronous* way. Analogously, signals may be transmitted in a *synchronous* way or *spooled*.

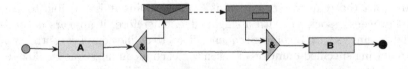

Fig. 7. Send and receive actions in a control flow

As sending and receiving are activities, they appear in control flows just as all other activities. Furthermore, sending and receiving (as activities) produce tokens when executed, and they may change the state in other ways. This has to be distinguished from the events that occur in connection with messaging.

The sending of a message creates an event SENT, and the receiving of a message creates an event RECEIVED. Besides these further message-related events can occur: DELAYED, LOST, REJECTED, UNDELIVERABLE, etc. It seems not advisable to create specific notation for each of these events, in particular not, if messaging events are integrated with other events using a temporal language for creating complex events.

4.1 Message Processing

A detailed ASM specification of communication actions (send, receive) has been done for S-BPM [11]. We adopt this specification in our modelling approach. In this model each receiver is equipped with an *inputPool*, which is subject to size restrictions concerning the total capacity (number of messages), the number of messages from a particular sender, the number of messages of a particular type, and the number of messages of a particular type coming from a particular sender. The bound 0 is used to indicate *synchronous communication*, otherwise it is *asynchronous*. Different strategies for handling violations to these bounds have been specified.

For sending and receiving the actor (sender or receiver) chooses between several alternatives defined by receiver and meassage type, and prepares a *messageToBeHandled*. For sending this is a *messageToBeSent*, for receiving this is an *expectedIncomingMessage*. Then *TryAlternative*$_{\text{commAct}}$ actually tries to send or receive the message.

Let us now sketch the ASM specification for message specification. Details can be found in the appendix of [11].

Perform (actor, CommAct, state) =
IF NonBlockingTryRound(actor, state) **THEN**
    IF TryRoundFinished(actor, state) **THEN**
       InitializeBlockingTryRounds (actor, state)
    ELSE *TryAlternative*$_{CommAct}$ (actor, state) **ENDIF**
ENDIF
IF BlockingTryRound(actor, state) **THEN**
    IF TryRoundFinished (actor, state) **THEN**
       InitializeRoundAlternatives (actor, state)
    ELSIF Timeout (actor, state, timeout (state)) **THEN**
       Interrupt$_{CommAct}$ (actor, state)
    ELSIF UserAbruption(actor, state) **THEN**
       Abrupt$_{CommAct}$ (actor, state)
    ELSE *TryAlternative*$_{CommAct}$ (actor, state) **ENDIF**
ENDIF
    with

$TryAlternative_{CommAct}$ (actor, state) =

$ChoosePrepareAlternative_{CommAct}$ (actor, state) **seq** $Try_{CommAct}$ (actor, state)

In a simple messaging model receivers are well-defined activities, i.e. the actors associated with the receiving activities. However, such a view is rather static. More generally, the state of the processes can be exploited to define receivers in a much more dynamic way. In particular, the receivers may depend on control, event, data and resource conditions just like flow nodes, and the dynamic determination of receivers can be part of the data operations.

4.2 Types of Events

There is a vast amount of possible event types: internal events, timing events, messaging events, etc. In general, events refer to something atomic that "has happened".

Internal events can refer to particular progress in the process flow, e.g. activities may be enabled, started, completed, but also postponed, interrupted, cancelled, delegated, delayed, etc., gateways may be waiting, enabled, fired, etc., and the whole process may have started or ended.

Likewise messages may have been sent, received, rejected, lost, retried (n'th time), etc., and signals may have been started, receiving, received, spooled, etc.

External events refer to activities that are happening in the environment, which can be captured by *monitored functions* in the ASM specification. Events can be combined to *complex events* using the usual logical junctors.

Events can also refer to *time*, for which a discrete, linear time model can be used. In distributed systems it is advisable to refer to local time (time depends on location) in order to avoid the problem of clock synchronisation. With time complex events can be created using temporal relationships: (directly) before, (directly) after, simultaneous, etc. Logically, time events and internal events can be combined.

4.3 Event Conditions and Actions

Thus, they main purpose of events is to enable the specification of *event conditions*, with which the firing of flow node instances becomes dependent on the progress of the process instance. Event conditions have been foreseen for all types of flow nodes, even starting or termination of a process (instance) can be made dependent on an event.

However, event conditions (as specified so far) refer to the firing of flow nodes. Nonetheless, event conditions can also be used to refine the control operation, e.g. deciding, on which outgoing sequence flows tokens shall be created. An example of the latter behaviour is captured by event-based gateways in BPMN. However, using events in the state permits more general definitions of complex gateways or event-driven activities.

Event actions have also been foreseen for all types of flow nodes. The purpose of event actions is to record those events that are relevant for the continuation of the process, as not all events are relevant to be recorded. That is, the recording would cover which event happened where and when, etc.

Some events may be disruptive requiring the process to be interrupted, (partially) rolled back and restarted with additional information. The handling of interrupts refers to exception handling.

5   Interaction and Data Handling

Interaction refers to the detailed specification, how basic activities are executed. The key idea is that each basic activity defines a *dialogue*, which may be broken down into several *dialogue steps*. We will first look into dialogue specifications, then briefly address how data conditions and actions can be specified. For the latter it is decisive that in each dialogue step certain data are consumed, while other data is produced. Here *data consumption* refers to the data that is needed (i.e. read) to perform any operation associated with the dialogue step, while *data production* refers to the data that is created by one of the operations associated with the dialogue step.

5.1   Interaction and Dialogues

Data consumption and production provide a local view on the data. Thus, both can be defined by small schemata associated with activities and then also dialogue steps. The integration of all these schemata defines a part of a global schema that underlies the data handling of the whole process for all instances.

Therefore, we distinguish between *database objects* defined by the global schema and *dialogue objects* defined by local schemata. Naturally, the local schemata define *views* on the global schema. By abstraction from individual activities we obtain a model of dialogue types, which we adopt from [28].

According to the previous discussion let us assume a *(global) database schema*. Elements appearing in instances of the global schema are referred to as *database objects*. These may be relations, trees, graphs, arrays, etc. We deliberately leave this open in order to stress that any data model (relational, nested, object-oriented, tree-based, graph-based, etc.) may be used here, but we must assume a *query language* that can be used to define *views*.

Then a *dialogue type* is basically defined by such a view plus operations. A *dialogue object* is an element in such a view for a particular database instance plus the operations of the dialogue type restricted to this element.

Based on the database schema operations (usually transactions) can be defined. Such *db-operations* require some input types and a specification of the actual updates of the database by means of db-programs.

Analogusly, operations on dialogue types can be defined. Such *dialogue-operations* also require input types plus a *selection type* defined on the view and a *body* that specifies which db-operations are used and which other dialogue objects are to be created.

Usually, on a high level of abstraction the concrete definition of the database schema, the queries defining the views and the db-operations are left abstract. According to our discussion a basic activity can be defined by a set of dialogue types defining implicitly data consumption and production and the flow of dialogue steps.

5.2 Data Conditions and Actions

The assumed global schema can be used to define data conditions for all flow nodes in the control flow specification. Such *data conditions* are expressed as Boolean queries. Analogously, the db-operations define *data actions* that may also be associated with the flow nodes in the control flow.

However, for activities the data actions result from the defining dialogue operations. Only the initialisation of the starting dialogue type has to be defined.

Data actions have global effects on the state. In particular, data conditions in remote parts of the specified process now depend on activities that seem to be independent. Thus, data conditions and actions require additional consistency verification due to these hidden dependencies. Furthermore, dialogue operations may also be defined directly on the views requiring a translation of the view updates to database updates. How to verify consistency with respect to data dependencies, is still a matter of research and will not be stressed further in this article.

6   Actors, Roles and Exceptions

The flow-centric specification of business processes emphasises the activities, their effects, and controlling conditions. Those who have to perform the (basic) activities are referred to as *actors*. Thus, to complete the picture, actors have to be associated with all activities. Then it is the *responsibility* of the actor to perform the activity using the associated dialogue objects as tools.

6.1 Responsibilities and Decision Making

Therefore, instead of directly associating actors, activities are usually assigned to *roles*, which are names representing an assortment of *obligations* and *rights*. In particular, some decisions on the flow of a process depend on data and events, while others have to be made explicitly by actors in certain roles.

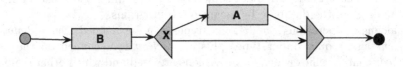

Fig. 8. Control flow with optional activity

Fig. 9. Control flow with explicitly marked optional activity

Example 7. In the control flow illustrated in Fig. 8 the activity **A** is optional. The question is who decides, if **A** is executed or not?

- Activity **B** may contain a data action such that the follow-on exclusive split can evaluate a data condition, which determines, if **A** or nothing is done. In this case, actually the actor associated with **B** is responsible for the decision.
- Alternatively, the data condition associated with the gateway can be defined elsewhere, if embedded in a larger control flow.
- We may associate a role also with the gateway, which, however, would blur the distinction between gateways that fire immediately when enabled and activities that usually depend on actor interaction.
- Finally, the decision may be a right of the actor associated with activity **A**, in which case the control flow is misleading.

In order to cope with situations as in Example 7 we have to distinguish cases, where decisions are based on pure data conditions from those, where it is the responsibility of the actor him/herself to decide about executing an optional activity. Therefore, deontic rules should be supported, some of which can be easily marked in control flows (see [20,22] for details).

Example 8. For instance, in Example 7 the activity **A** could be marked as optional as illustrated in Fig. 9. Executing **O(A)** means that the actor associated with this activity either executes **A** or nothing. Analogously, alternatives and forbidden actions can be handled this way [20,22].

A more fine-grained specification of rights and obligations associated with roles can be obtained by *deontic action logic*. The *atoms* of the logic are defined as follows:

- $do(r, a)$ means that an actor in role r *executes* action a.
- $\mathbf{P}do(r, a)$ means that an actor in role r is *permitted* to execute action a.
- $\mathbf{F}do(r, a)$ means it is *forbidden* for an actor in role r to execute action a.
- $\mathbf{O}do(r, a)$ means that an actor in role r is *obliged* to execute action a.

Formulae in the logic are constructed from the atoms with the usual logical junctors. Then deontic constraints give rise to another verification task, i.e. to check if processes can be executed under the given constraints.

6.2 Exception Handling

An *exception* is a disruptive event, i.e. an exception causes that the running process instance is interrupted regardless of its state, a complete or partial roll-back to a consistent state will be executed, and depending on the kind of the exception a restart from a consistent state with a different continuation process will be launched. Examples of exceptions are order cancellation, bancrupcy of a customer, serious breakdown, etc.

The rollback may affect a single or a few activities, but also the complete process may have to be rolled back. The rollback triggered by an exception is a process in its own right. It will first collect activities that have to be undone, either perform **undo** operations as in databases or apply compensation activities if possible. **Undo** operations may require adequate data, i.e. the detailed specifications of activities, gateways, decisions, etc. have to be refined.

At the end of the rollback the process should have reached a consistent previous state, from which the process instance will resume, i.e. continue in the modified state, which also contains information about the exception.

Continuation processes can be part of the process specification. Thus, we may use conservative extensions to specify alternative paths in case of particular exceptions:

> **IF** notInterrupted **THEN** $\langle$as specified$\rangle$
> **ELSIF** exception$_1$ **THEN** continuation$_1$ ...

Usually, continuation processes give rise to subprocesses.

Exception handling is still subject to research, so we dispense with discussing further details in this article.

7   Conclusions

In this article we briefly outlined business process specifications grounded in horizontal refinements, which enables the integration of several sub-systems addressing control flow, message handling, event handling, data handling, exception handling, etc. While some of the work has reached already a mature state, other parts are still under investigation. The final goal is to complete the specification of an integrated businss process model H-BPM with unambiguous semantics defined by ASMs.

The main emphasis of H-BPM is to capture all relevant aspects of BPM and to formally define the semantics using Abstract State Machines in order to enable validation and formal verification. Usually actors and data handling have been neglected, while other constructs handled superficially (see [16] for a discussion of major parts of BPMN 2.0). In this way H-BPM aims to enable seamless modelling of business processes on all levels of abstraction as well as horizontal and vertical model integration by means of formal refinement.

Regarding vertical integration is achieved by further refining the involved ASMs in a development process that is targeting the executable specification of a workflow engine that is enriched with components for data and dialogue handling and exception processing. Throughout the process rigorous quality assurance methods have to be applied.

In this brief article we only gave an overview of the most relevant aspects of H-BPM without going too much into details. In particular, several extensions were not yet handled such as modularity by subprocesses with replacement semantics, which will have many implications on the constructs introduced so far, further details of the ASM specifications, multiple view presentations, in particular coarse and fine presentation of specifications, etc. We also did not stress how to handle validation nor how to effectively perform verification.

Regarding our future research further extensions are envisioned. These comprise the possibility to postpone or cancel activities with the corresponding meta-level rights and obligations. Furthermore, we intend to investigate adaptivity in business process specifications, in particular with respect to preferences, exception handling and ad-hoc changes. Adaptivity is the subject of a running research project AdaBPM.

References

1. Abramowicz, W., Filipowska, A., Kaczmarek, M., Kaczmarek, T.: Semantically enhanced business process modelling notation. In: Hepp, M., et al. (eds.) S-BPM. CEUR Workshop Proceedings, vol. 251 (2007). CEUR-WS.org
2. Blass, A., Gurevich, Y.: Abstract state machines capture parallel algorithms. ACM Trans. Comput. Log. 4(4), 578–651 (2003)
3. Blass, A., Gurevich, Y.: Abstract State Machines capture parallel algorithms: Correction and extension. ACM Trans. Comput. Log. 9(3) (2008)
4. Börger, E.: Approaches to modeling business processes: a critical analysis of BPMN, workflow patterns and YAWL. Softw. Syst. Model. 11(3), 305–318 (2012)
5. Börger, E., Sörensen, O.: BPMN core modeling concepts: inheritance-based execution semantics. In: Embley, D., Thalheim, B. (eds.) Handbook of Conceptual Modeling: Theory, Practice and Research Challenges, pp. 287–335. Springer, Heidelberg (2011)
6. Börger, E., Sörensen, O., Thalheim, B.: On defining the behavior of OR-joins in business process models. J. Univ. Comput. Sci. 15(1), 3–32 (2009)
7. Börger, E., Stärk, R.: Abstract State Machines. Springer, Heidelberg (2003)
8. Börger, E., Thalheim, B.: A Method for verifiable and validatable business process modeling. In: Börger, E., Cisternino, A. (eds.) Advances in Software Engineering. LNCS, vol. 5316, pp. 59–115. Springer, Heidelberg (2008)
9. Börger, E., Thalheim, B.: Modeling workflows, interaction patterns, web services and business processes: The ASM-based approach. In: Börger, E., Butler, M., Bowen, J.P., Boca, P. (eds.) ABZ 2008. LNCS, vol. 5238, pp. 24–38. Springer, Heidelberg (2008)
10. Dumas, M., La Rosa, M., Mendling, J., Reijers, H.A.: Fundamentals of Business Process Management. Springer, Heidelberg (2013)
11. Fleischmann, A., et al.: Subject-Oriented Business Process Management. Springer, Heidelberg (2012)
12. Geist, V.: Integrated Executable Business Process and Dialogue Specification. Ph.D. thesis, Johannes Kepler University Linz, Austria (2011)
13. Gurevich, Y.: Sequential Abstract State Machines capture sequential algorithms. ACM Trans. Computat. Log. 1(1), 77–111 (2000)
14. Kopetzky, T., Geist, V.: Workflow charts and their precise semantics using abstract state machines. In: Rinderle-Ma, S., Weske, M. (eds.) Proceedings of EMISA 2012 - Der Mensch im Zentrum der Modellierung, Vienna, Austria (2012). LNI, pp. 11–24. Kllen-Verlag, Bonn (2012)

15. Kossak, F., Illibauer, C., Geist, V.: Event-based gateways: open questions and inconsistencies. In: Mendling, J., Weidlich, M. (eds.) BPMN 2012. LNBIP, vol. 125, pp. 53 67. Springer, Heidelberg (2012)

16. Kossak, F., et al.: A Rigorous Semantics for BPMN 2.0 Process Diagrams. Springer (2014, forthcoming)

17. Natschläger, C.: Deontic BPMN. In: Hameurlain, A., Liddle, S.W., Schewe, K.-D., Zhou, X. (eds.) DEXA 2011, Part II. LNCS, vol. 6861, pp. 264–278. Springer, Heidelberg (2011)

18. Natschläger, C.: Towards a BPMN 2.0 ontology. In: Dijkman, R., Hofstetter, J., Koehler, J. (eds.) BPMN 2011. LNBIP, vol. 95, pp. 1–15. Springer, Heidelberg (2011)

19. Natschläger, C., Geist, V., Kossak, F., Freudenthaler, B.: Optional activities in process flows. In: Rinderle-Ma, S., Weske, M. (eds.) Der Mensch im Zentrum der Modellierung. LNI. Kllen-Verlag, Bonn (2012)

20. Natschläger, C., Kossak, F., Schewe, K.D.: BPMN to Deontic BPMN: A trusted model transformation. Journal of Software and Systems Modelling (2014, to appear)

21. Natschläger, C., Schewe, K.D.: A flattening approach for attributed type graphs with inheritance in algebraic graph transformation. Electron. Commun. EASST **47**, 160–173 (2012)

22. Natschläger-Carpella, C.: Extending BPMN with Deontic Logic. Logos Verlag, Berlin (2012)

23. Petri, C.A.: Communication with automata. Ph.D. thesis, Universität Hamburg (1966)

24. Recker, J.C., Rosemann, M., Indulska, M., Green, P.: Business process modeling: A comparative analysis (2009)

25. Reichert, M., Weber, B.: Enabling Flexibility in Process-Aware Information Systems: Challenges, Methods, Technologies. Springer, Heidelberg (2012)

26. Scheer, A.W.: ARIS - Business Process Modeling. Springer, Heidelberg (2000)

27. Schewe, K.-D.: Horizontal and vertical business process model integration. In: Decker, H., Lhotská, L., Link, S., Basl, J., Tjoa, A.M. (eds.) DEXA 2013, Part I. LNCS, vol. 8055, pp. 1–3. Springer, Heidelberg (2013)

28. Schewe, K.D., Schewe, B.: Integrating database and dialogue design. Knowl. Inf. Syst. **2**(1), 1–32 (2000)

29. ter Hofstede, A.M., et al. (eds.): Modern Business Process Automation: YAWL and its Support Environment. Springer, Heidelberg (2010)

30. Weske, M.: Business Process Management: Concepts, Languages, Architectures. Springer, Heidelberg (2012)

31. Wohed, P., van der Aalst, W.M.P., Dumas, M., ter Hofstede, A.H.M., Russell, N.: On the suitability of BPMN for business process modelling. In: Dustdar, S., Fiadeiro, J.L., Sheth, A.P. (eds.) BPM 2006. LNCS, vol. 4102, pp. 161–176. Springer, Heidelberg (2006)

32. Wong, P.Y.H., Gibbons, J.: A process semantics for BPMN. In: Liu, S., Araki, K. (eds.) ICFEM 2008. LNCS, vol. 5256, pp. 355–374. Springer, Heidelberg (2008)

33. zur Muehlen, M., Recker, J.C., Indulska, M.: Sometimes less is more: are process modeling languages overly complex? In: Taveter, K., Gasevic, D. (eds.) 3rd International Workshop on Vocabularies, Ontologies and Rules for the Enterprise. IEEE, Annapolis (2007)

Exact and Approximate Generic Multi-criteria Top-k Query Processing

Mehdi Badr[1,2] and Dan Vodislav[1(✉)]

[1] ETIS, ENSEA - University of Cergy-Pontoise - CNRS, Cergy, France
dan.vodislav@u-cergy.fr
[2] TRIMANE, 57 rue de Mareil, Saint-Germain-en-Laye, France
Badr.Mehdi@gmail.com

Abstract. Many algorithms for multi-criteria top-k query processing with ranking predicates have been proposed, but little effort has been directed toward genericity, i.e. supporting any type of access to the lists of predicate scores (sorted and/or random), or any access cost settings. In this paper we propose a general approach to exact and approximate generic top-k processing. To this end, we propose a general framework (GF) for generic top-k processing, able to express any top-k algorithm and present within this framework a first comparison between generic algorithms. In previous work, we proposed BreadthRefine (BR), a generic algorithm that considers the current top-k candidates as a whole instead of focusing on the best candidate for score refinement, then we compared it with specific top-k algorithms. In this paper, we propose two variants of existing generic strategies and experimentally compare them with the BR breadth-first strategy, showing that BR leads to better execution costs. We also extend the notion of θ-approximation to the GF framework and present a first experimental study of the approximation potential of top-k algorithms on early stopping.

Keywords: Top-k query processing · Ranking · Multi-criteria information retrieval

1 Introduction

We address the problem of top-k multi-criteria query processing, where queries are composed of a set of ranking predicates, each one expressing a measure of similarity between data objects on some specific criterion. Unlike traditional Boolean predicates, similarity predicates return a relevance score in a given interval. The query also specifies an aggregation function that combines the scores produced by the similarity predicate of each criterion. Query results are ranked following the global score and only the best k ones are returned.

Ranking predicates acquired an increasing importance in today's data retrieval applications, especially with the introduction of new, weakly structured data types: text, images, maps, etc. Searching such data requires content-based information retrieval (CBIR) techniques, based on predicates measuring the similarity

© Springer-Verlag Berlin Heidelberg 2015
A. Hameurlain et al. (Eds.): TLDKS XVIII, LNCS 8980, pp. 53–79, 2015.
DOI: 10.1007/978-3-662-46485-4_3

select * from Object o　　　　　　　**select * from** Monument m
order by $\mathcal{F}(p_1(o), ..., p_m(o))$　　　**order by** $near(m.$address, here$())$ +
limit k　　　　　　　　　　　　　　　$similar(m.$photo, myPhoto$)$ +
　　　　　　　　　　　　　　　　　　　$ftcontains(m.$descr, 'Renaissance sculpture'$)$
　　　　　　　　　　　　　　limit 1

Fig. 1. General form and example of a top-k query

between data objects, by using content descriptors such as keyword sets, image descriptors, geographical coordinates, etc. We consider here the case of *expensive* ranking predicates over data objects, whose specificity is that their evaluation cost dominates the cost of the other query processing operations.

The general form of the top-k queries that we consider is expressed in Fig. 1. The query asks for the k best objects following the scores produced by m ranking predicates $p_1, ..., p_m$, aggregated by a monotone function $\mathcal{F}$.

Figure 1 also presents a query example coming e.g. from a smartphone touristic application, where the visitor of a Renaissance monument, after finishing the visit, searches for another similar monument (the "best" one) on three criteria: *near* to his current location, containing a *similar detail* to some picture taken with the smartphone, and exposing *Renaissance artworks, preferably sculptures*. Here, the aggregate function is a simple sum.

As in this example, expensive ranking predicates come often from the evaluation of similarity between images, text, locations and other multimedia types, whose content is described by numerical vectors. This results in expensive searches in highly dimensional spaces, based often on specific multidimensional index structures [3]. Note that most of the ranked predicates in this case come from binary predicates $sim(o_1, o_2)$ evaluating similarity between objects, transformed into unary ranked predicates $p(o) = sim(o, q)$ evaluating the similarity with a query object q.

In many cases, predicates are evaluated by distant, specialized sites, that provide specific web services, e.g. map services evaluating spatial proximity, photo sharing sites allowing search of similar images, specialized web sites proposing rankings for hotels, restaurants, etc. Internet access to such services results into expensive predicate evaluation by distant, independent sites. Moreover, the control over predicate evaluation is minimal most of the time, reduced to the call of the provided web service.

For each query, a ranking predicate may produce a score for each object. Following, we call *a source* the collection of scores produced by a ranking predicate for the set of data objects. The list of scores may be produced e.g. by accessing a local index structure that returns results by order of relevance. We consider here the general case, where the access to the scores of a source is limited to sorted and/or random access. This allows three possible types for a source S:

– *S-source*: sorted access only, through the operator *getNext(S)* returning the pair (o, s) containing the identifier o of the object with the next highest score s.
– *R-source*: random access only, through the operator *getScore(S, o)* returning the score of a given object o.
– *SR-source*: a source with both sorted and random access.

S_1 (S)	S_2 (SR)	S_3 (R)
$(o_2, 0.4)$	$(o_3, 0.9)$	$(o_1, 0.9)$
$(o_1, 0.3)$	$(o_1, 0.2)$	$(o_2, 0.7)$
$(o_4, 0.25)$	$(o_4, 0.15)$	$(o_3, 0.8)$
$(o_3, 0.2)$	$(o_2, 0.1)$	$(o_4, 0.6)$

Access	Retrieved	candidates	U_{unseen}
		$\emptyset$	3.0
S_1/S	$(o_2, 0.4)$	$\{(o_2, [0.4, 2.4])\}$	2.4
S_2/S	$(o_3, 0.9)$	$\{(o_2, [0.4, 2.3]), (o_3, [0.9, 2.3])\}$	2.3
S_2/R	$(o_2, 0.1)$	$\{(o_2, [0.5, 1.5]), (o_3, [0.9, 2.3])\}$	2.3
S_3/R	$(o_3, 0.8)$	$\{(o_2, [0.5, 1.5]), (o_3, [1.7, 2.1])\}$	2.3
S_2/S	$(o_1, 0.2)$	$\{(o_2, [0.5, 1.5]), (o_3, [1.7, 2.1]), (o_1, [0.2, 1.6])\}$	1.6

Fig. 2. Examples of sources and query execution for the query example

The general idea of a top-k algorithm is to avoid computing all the global scores, by maintaining a list of candidate objects and the interval $[L, U]$ of possible global scores for each of them. The initial interval of a candidate is obtained by aggregating the minimum/maximum source scores.

The monotonicity of the aggregation function ensures that further source accesses always decrease the upper bound U and increase the lower bound L. The algorithm stops when the score of the best k candidates cannot be exceeded by the other objects anymore.

Figure 2 presents a possible execution for the example query in Fig. 1. We suppose S_1 is an S-source, S_2 an SR-source, S_3 an R-source; scores are presented in descending order for S/SR sources and by object identifier for R-sources. Local scores belong to the $[0, 1]$ interval in this example, so the initial global score interval is $[0, 3]$ for all objects.

We note *candidates* the set of candidates and U_{unseen} the maximum score of unseen objects (not yet discovered in some source). Initially, *candidates* $= \emptyset$ and $U_{unseen} = 3$.

- A sorted access to S_1 retrieves $(o_2, 0.4)$, so o_2's global score interval becomes $[0.4, 2.4]$. Also U_{unseen} becomes 2.4 because further scores in S_1 cannot exceed 0.4.
- Then, a sorted access to S_2 retrieves $(o_3, 0.9)$. This adds a new candidate (o_3), lowers U_{unseen} to 2.3 (further S_2 scores cannot exceed 0.9), but also lowers the upper bound of o_2 to 2.3, because the maximum score of S_2 is now 0.9.
- Next, a random access to S_2 for o_2 retrieves $(o_2, 0.1)$. This changes only the global score interval of o_2.
- A random access to S_3 for o_3 retrieves $(o_3, 0.8)$ and changes the global score interval of o3.
- A sorted access to S_2 retrieves $(o_1, 0.2)$. This adds a new candidate (o_1), lowers U_{unseen} to 1.6, but does not lower the maximal global score of the other candidates because o_2 and o_3 already know their S_2 scores.

The minimum global score of o_3 exceeds now both U_{unseen} and the maximum global score of all the other candidates and the execution stops since o_3 is surely the best (top-1) object.

2   Related Work and Contribution

A large spectrum of top-k query processing techniques [11] has been proposed at different levels: query model, access types, implementation structures, etc. We consider here the most general case, of simple top-k selection queries, with expensive access to sources, limited to individual sorted/random probes, without additional information about local source scores/objects, and out of the database engine.

This excludes from our context join queries [10,17] or interaction with other database operators for query optimization [10,12,13]. We consider sequential access only, parallel processing is out of the scope of this paper. We exclude also approaches such as TPUT [5], KLEE [16] or BPA [1], able to get several items at once, or having statistical information available about scores, or having also the local rank. Algorithms such as LARA [14], that optimize the management of the candidate list, are orthogonal to our approach for expensive predicates, which focuses on source access.

In this context, top-k algorithms proposed so far fit with the general method presented in the example of Fig. 2 and propose their own heuristic for deciding the next access to a score source. However, most algorithms focus on specific source types and cost settings.

Algorithms such as *NRA* [7] (No Random Access) and *StreamCombine* [9] consider only S-sources. NRA successively consults all the sources in a fixed order, while StreamCombine selects at each step the next access based on a notion of *source benefit*.

Other algorithms consider only SR-sources. The best known is *TA* [7] (Threshold Algorithm), which consults sorted sources in a fixed order (like NRA), but fully evaluates the global score of each candidate through random access to the other sources. The algorithm stops when at least k global scores exceed U_{unseen}. Among the extensions of TA we cite *QuickCombine* [8], which uses the same idea as StreamCombine to select the next sorted source to probe, and *TAz* [4], which considers an additional set of R-sources besides the SR-sources. *CA* [7] (Combined Algorithm) is a combination of TA with NRA that considers random accesses being h times more expensive than sorted ones. It reduces the number of random probes by performing h sorted accesses in each source before a complete evaluation of the best candidate by random probes.

Also supposing cost asymmetry, a third category of algorithms considers one cheap S-source (providing candidates) and several expensive R-sources. *Upper* [4,15] focuses on the candidate with the highest upper bound U and performs a random probe for it, unless $U < U_{unseen}$, in which case a sorted access is done. The choice of the next R-source to probe is based on a notion of source benefit, dynamically computed. *MPro* [6] is similar to Upper, but fixes for all the candidates the same order for probing the R-sources, determined by sampling optimization.

Surprisingly, little effort has been made towards generic top-k processing, i.e. adapted to any combination of source types and any cost settings. To our knowledge, besides our BreadthRefine proposal described below, *NC* [19] (Necessary Choices) is the only other generic approach, however limited to the case of results *with complete scoring*. NC proposes a framework for generic top-k algorithms, a strategy SR that favors sorted accesses, and a specific algorithm SR/G that uses sampling optimization to find the parameters that produce the best fit given the source settings.

Approximate top-k processing has been considered in several approaches, the most usual one being the approximation by early stopping, i.e. considering the current top-k objects at some point during the execution as an approximate result. Since early stopping comes with no guarantees on the quality of the result, several constraints providing such guarantees have been considered. For instance, a variant of the TA algorithm, called TA$_\theta$ [7], defines an approximation parameter $\theta > 1$ and the θ-approximation of the top-k result as being a set K_a of k objects such that $\forall x \in K_a, \forall y \notin K_a, \theta \times score(x) \geq score(y)$ (global and local scores are considered to belong to the [0,1] interval). The intuition behind this condition is that the ratio between the score of the best missed object in the approximate result (best false negative) and that of the worst false positive cannot exceed θ. To obtain a θ-approximation, TA$_\theta$ simply changes the threshold condition: the algorithm stops when at least k objects have a global score $\geq U_{unseen}/\theta$, i.e. TA$_\theta$ is equivalent to an early stopping of the TA algorithm.

Other approximation algorithms for top-k selection queries are proposed in [18], for S-source algorithms, or in the *KLEE* system [16] for top-k processing in distributed environments. Note that [18] is based on dropping candidates that have low probability to be in the top-k and provides probabilistic guarantees for the result, but requires knowledge about score distribution in sources.

In previous work, we have proposed *BR* (BreadthRefine) [2], a generic algorithm that uses a breadth-first strategy for top-k processing in a larger context than NC, i.e. with incomplete scoring. The BR strategy considers the current top-k as a whole to be refined, while all the other proposed strategies focus on the best candidate. BR has been compared to algorithms of the three categories mentioned above and proved that it successfully adapts to their specific settings, with better cost.

In this paper, we address exact and approximate multi-criteria top-k query processing at a general level, proposing generalizations of existing algorithms to the generic case and aiming at a comparison of algorithm strategies. More precisely, our contributions are the following:

– A general framework *GF* for generic top-k multi-criteria query processing, that allows expressing any top-k algorithm of our context, thus providing a basis for comparative analysis and generalization.
– The BR algorithm is generic (adapted to any combination of source types and any cost settings), but it was only compared to specific top-k algorithms, since the only other generic approach, introduced by NC, is hardly comparable with BR. As further detailed in Sect. 3, the difficulty to compare with NC

comes mainly from the fact that, unlike BR and most top-k algorithms, NC is not fully heuristic and strongly depends on a sampling optimization phase. We propose here new, comparable generic variants of the BR, NC and CA algorithms and experimentally compare these generic strategies, showing that BR leads to better execution costs.

- A generalization of θ-approximation computing in the context of GF, and a first experimental study of the ability of top-k multi-criteria algorithms to produce good approximate results on early stopping, showing that the BR strategy comes with a better approximation potential.

We do not directly address here algorithm optimality issues. Fagin *et al.* demonstrate in [7] that NRA and TA algorithms are instance optimal, i.e. for any database instance, no top-k algorithm can improve the execution cost with more than a constant factor. They also show that algorithms based on a dynamic choice of the next source to access (such as BR or, more generally, algorithms expressed in GF) may not be instance optimal, although they may have in practice better execution costs. Even if BR and the other generic algorithms we consider are not instance optimal, our goal is to experimentally demonstrate that the BR strategy leads to better performances. Note however that, as shown in [7], BR could be adapted to become instance optimal by adding source accesses that guarantee every source to be accessed at least once every C steps, for some constant C.

The rest of the paper is organized as follows: the next section introduces the generic framework for top-k multi-criteria processing, then proposes and compares in this context new generic variants for BR, NC and CA. Section 4 presents our approach for top-k approximation in the general framework, then we report experimental results and end with conclusions.

3   Generic Top-k Framework and Algorithms

We propose *GF*, a *generic framework* for multi-criteria top-k processing (Fig. 3). GF provides a common, general form for expressing any top-k algorithm in our context. It facilitates comparison between top-k algorithms and strategies expressed in this common form. For instance, we benefit here from this common framework in the description of new variants of existing algorithms (NC and CA), compared then with our BR approach. Another major benefit of GF is that new properties expressed and proved in this general framework are true for any top-k algorithm - for instance, the θ-approximation properties presented in Sect. 4.

As in the example of Fig. 2, GF considers a top-k algorithm as a sequence of accesses to the sources, that progressively discover scoring information about data objects. The input parameters are the query q and the set of sources S. Query q specifies the number k of results to return and the monotone aggregation function $\mathcal{F}$, while the set of sources S materializes the scores returned by the query's ranking predicates.

GF $(q, \mathcal{S})$
    candidates $\leftarrow \emptyset$; $U_{unseen} \leftarrow \mathcal{F}(max_1, ..., max_m)$; *... //other local variables*
    repeat *//choice between sorted or random access*
       if *SortedAccessCondition()* **then** *//sorted access*
          $S_j \leftarrow$ **BestSortedSource()** *//choice of a sorted source*
          $(o, s) \leftarrow$ getNext(S_j) *//sorted access to the selected source*
          Update *candidates*, U_{unseen} and other local variables
       else *//random access*
          $c \leftarrow$ **ChooseCandidate()** *//choice of a candidate*
          $S_j \leftarrow$ **BestRandomSource**(c) *//choice of a random source*
          $s \leftarrow$ getScore(S_j, c) *//random access to the selected source*
          Update *candidates* and other local variables
       endif
    until *StopCondition()*
    return *candidates*

Fig. 3. The GF generic top-k framework

In GF, algorithms maintain *a set of candidates* (initially empty) with their interval of scores, *the threshold* U_{unseen} (initialized with the aggregation of the maximum scores max_j of the sources), and possibly other local data structures.

Notations

- For a candidate c, we note $[L(c), U(c)]$ its current interval of scores.
- We note $\mathcal{U}_k$ (respectively $\mathcal{L}_k$) the current subset of k candidates with the best k upper (lower) bound scores[1].
- We note U_k the current k-th highest upper bound score among the candidates, i.e. $U_k = min_{c \in \mathcal{U}_k}(U(c))$, respectively L_k the current k-th highest lower bound score, $L_k = min_{c \in \mathcal{L}_k}(L(c))$.
- We note $\chi \in \mathcal{U}_1$ the candidate with the current best upper bound score.

Note that the monotonicity of the aggregation function guarantees that the threshold and the upper bound of candidate scores only decrease, while their lower bound only increase during the algorithm.

One source access is performed at each iteration, the access type being decided by the *SortedAccessCondition* predicate. In the case of a sorted access, a source S_j is chosen by the **BestSortedSource** function, then is accessed through *getNext*. The returned object-score couple is used to update the threshold, the set of candidates and the local variables. The retrieved object is added/updated in the *candidates* set and objects not yet retrieved in S_j update their upper bounds.

Update also includes *the discarding of non-viable candidates*. A candidate c with $U(c) < L_k$ is called non-viable because it will never be in the top-k result since at least k candidates surely have better scores.

In the case of a random access, the **ChooseCandidate** function selects a candidate c, then **BestRandomSource** gives a random source to probe for it.

---
[1] With random selection among candidates with the same score if necessary.

After the random access through *getScore*, the *candidates* set and local variables are updated (among candidates, only c changes).

The end of the algorithm is controlled by the generic *StopCondition* predicate, which depends on the type of top-k result expected (e.g. with complete or incomplete scoring). The earliest end is obtained with predicate

$$StopCondition \equiv (|candidates| = k \wedge L_k \geq U_{unseen}) \tag{1}$$

i.e. only k candidates are viable and there is no viable unseen object. Since this result may have incomplete scoring, additional conditions are necessary to ensure properties such as ordering or complete scoring of the results.

It is simple to demonstrate that this condition is necessary and sufficient for a correct top-k result. Sufficiency is trivial, the k remaining candidates form a correct top-k, because their scores are at least L_k, while the score of non viable candidates and that of unseen objects is $\leq L_k$. Necessity comes from the fact that if condition (1) is not true, either $|candidates| < k$ (and then we do not have k candidates to form the result), or $|candidates| > k$ (and then any of the viable candidates still may have a final score that corresponds to a top-k object), or $L_k < U_{unseen}$ (and then an unseen object may belong to the top-k).

It is easy to see that any top-k algorithm in our context can be expressed in GF. Indeed, for a given query and set of sources, each algorithm is equivalent to the sequence of accesses to the sources it produces, which can be obtained with a sequence of decisions about the access type, the source and the candidate for random probes.

Note that this is not true for instance with the NC framework [19], in which one chooses first a candidate among the k highest upper bound scores, then a source in which the candidate has not been yet retrieved. This is not compatible with algorithms that fix the order of accessing sources, such as NRA: a source in which candidates with the current k highest upper bound scores have been already found cannot be selected for the next step.

As an example, the NRA algorithm can be expressed in GF with *SortedAccessCondition* $\equiv$ *true* (only S-sources), a local variable keeping the last accessed source and a function **BestSortedSource** returning the next source in a round robin order.

Note that algorithms with SR-sources only, like TA, may avoid maintaining interval scores, because the global score of each candidate is immediately computed; however, this optimization is not relevant in our context, where cost is given by the access to the sources and not by the updates to local data structures.

Given its ability to express any top-k algorithm, the GF framework is a valuable tool for comparing top-k strategies. Following, we express in GF and compare three generic algorithms: a new variant of BR and new, generic and comparable variants of the NC and CA algorithms.

3.1   BreadthRefine

BreadthRefine (BR) [2] proposes a generic algorithm framework that can be instantiated to several variants. The main idea of the BR strategy is to maintain the set of current top-k candidates $\mathcal{U}_k$ as a whole, instead of focusing on the best candidate χ, which is the common approach.

BR was successfully compared with state of the art non-generic algorithms in their specific settings. We complete here this comparison by considering also two other generic top-k strategies, adapted for that purpose to our context.

The BR framework can be expressed in the more general GF framework by instantiating *SortedAccessCondition* and **ChooseCandidate** to realize the BR strategy.

- *SortedAccessCondition* $\equiv$ ($|candidates| < k$ **or** $U_{unseen} > U_k$ **or** *CostCondition()*).

  The *SortedAccessCondition* in the BR strategy combines three conditions: a sorted access is scheduled if (i) there are not yet k candidates, or (ii) an unseen object could belong to the current top-k $\mathcal{U}_k$ ($U_{unseen} > U_k$), or (iii) a generic *CostCondition* favors sorted access in the typical case where a random access is more expensive than a sorted one.

  Condition (ii) targets the decrease of U_{unseen} through sorted accesses and is the heart of the BR strategy for sorted sources. The common strategy for sorted access focuses only on the best candidate χ, and to be sure that χ (and not some unseen object) has the best upper score, a sorted access is scheduled if $U_{unseen} > U(\chi)$ to decrease U_{unseen} below $U(\chi)$. The BR strategy focuses on the whole current top-k: it maintains the whole $\mathcal{U}_k$ free of unseen objects, by scheduling a sorted access if $U_{unseen} > U_k$.
- The BR strategy is completed by the **ChooseCandidate** function for refinement by random probes. All the existing algorithms facing this choice systematically select the best current candidate χ. Instead, the BR strategy maintains the k best candidates as a whole by first selecting the least refined candidate in $\mathcal{U}_k$.

BR considers top-k with incomplete scoring, thus *StopCondition* is given by (1).

BR-Cost*. Several instantiations of the BR framework have been proposed in [2]. The one producing the best costs is *BR-Cost*, that fully implements the BR strategy and uses a *CostCondition* inspired from CA: if r is the ratio between the average costs of random and sorted accesses, then successive random probes must be separated by at least r sorted accesses.

In BR-Cost, **BestSortedSource** and **BestRandomSource** adopt a benefit-oriented strategy, inspired by StreamCombine [9] for choosing a sorted source, or by algorithms with controlled random probes such as Upper [4] for random access.

- For a sorted access, the benefit of source S_j is $Bs_j = (\partial \mathcal{F}/\partial S_j) \times N_j \times \delta_j/C_s(S_j)$, where $(\partial \mathcal{F}/\partial S_j)$ is the weight of S_j in the aggregation function, N_j the number of candidates in $\mathcal{U}_k$ not yet seen in S_j, δ_j the expected decrease of the score in S_j and $C_s(S_j)$ the cost of a sorted access in S_j. Since $(\partial \mathcal{F}/\partial S_j)$ cannot be computed for any monotone function $\mathcal{F}$, we consider here, for simplicity, *only the case of weighted sum*, in which $(\partial \mathcal{F}/\partial S_j) = coef_j > 0$, where

$coef_j$ is the coefficient corresponding to source S_j in the weighted sum. The value of δ_j can be obtained, e.g. by making one access ahead, with negligible extra cost.

The intuition behind this formula is that the benefit measures the potential refinement of the candidates score intervals, relative to the access cost. The sorted access to S_j refines not only the score interval of the retrieved object, but also that of objects not yet found in S_j; for these objects the upper bound decreases by $coef_j \times \delta_j$. This formula, borrowed from StreamCombine [9], only considers the N_j candidates of the current top-k not yet found in S_j.

- For a random access, the benefit of source S_j is $Br_j = coef_j \times (crtmax_j - min_j)/C_r(S_j)$, where $crtmax_j$ and min_j are respectively the current maximum score and the minimum score in S_j and $C_r(S_j)$ is the cost of a random probe in S_j. Note that $crtmax_j$ decreases in SR-sources (after sorted accesses), but remains constant (equal to max_j) in R-sources. Note also that $coef_j \times (crtmax_j - min_j)$ measures the reduction of the candidate's score interval size after a random probe in S_j, i.e. here also, the benefit expresses the refinement of the score interval of the accessed object, relative to the access cost.

We propose here **BR-Cost***, an improved variant of BR-Cost, using a different method for estimating r, in this case as a *ratio of benefits* instead of a ratio of access costs. We measure r as the ratio between the average benefit of making a sorted access vs a random one.

As for **BestSortedSource** and **BestRandomSource**, we consider the benefit of an access to a source as being related to its impact on the evolution toward the final top-k, measured by the decrease of the size of the interval of scores of the candidates. More precisely, the benefit of an access to a source is defined as the ratio between the refinement produced on *all* the candidate score intervals and the cost of that access.

Note that this definition corresponds to that used by **BestRandomSource** for random access, because only one candidate is impacted by a random probe, but generalizes the benefit used by **BestSortedSource**, by considering the decrease of score intervals for all the candidates, not only for those of the current top-k. This corresponds to an uniform model for the benefit of accessing any type of source and is more adapted for computing an average benefit. This approach also favors the comparison with the NC strategy.

Consider the case of a S-source S_j in Fig. 4 at the moment when the current score is $crtmax_j$ and Nr_j objects have not been yet accessed. A sorted access to S_j refines the score of the retrieved object, but also produces a decrease δ_j of $crtmax_j$ that affects the upper bound of the remaining $Nr_j - 1$ objects. For the retrieved object, the width of the score interval decreases with $coef_j \times (crtmax_j - min_j)$. For each one of the remaining $Nr_j - 1$ objects, the upper bound decreases with $coef_j \times \delta_j$.

In conclusion, the benefit of a sorted access to S_j is:

$$B_s(S_j) = coef_j \times (crtmax_j - min_j + (Nr_j - 1) \times \delta_j)/C_s(S_j) \qquad (2)$$

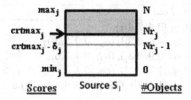

Fig. 4. Scores in a sorted source S_j

Benefit varies in time; if δ_j does not vary much, benefit globally decreases because $crtmax_j$ and Nr_j decrease. We approximate the average benefit by considering $\delta_j \approx (max_j - min_j)/N$, $crtmax_j \approx (max_j - min_j)/2$ and $Nr_j \approx N/2$:

$$\overline{B_s}(S_j) \approx coef_j \times (max_j - min_j)/C_s(S_j) \tag{3}$$

Note that the instant benefit $B_s(S_j)$ may also be computed at any moment if the total number of objects in the database is known or can be estimated. The instant benefit could be used e.g. as an alternative value for **BestSortedSource**, or for computing a variable ratio r in the BR-Cost* algorithm. Following, we only consider a fixed ratio r, based on the average source benefit.

Benefit for a random access is computed in a similar way, but in this case only the score interval of the selected candidate changes. If S_j is a SR-source, the benefit, respectively the average benefit of a random access are:

$$B_{rs}(S_j) = coef_j \times (crtmax_j - min_j)/C_r(S_j) \tag{4}$$

$$\overline{B_{rs}}(S_j) \approx coef_j \times (max_j - min_j)/2C_r(S_j) \tag{5}$$

For a R-source $crtmax_j = max_j$ all the time, therefore

$$B_r(S_j) = \overline{B_r}(S_j) = coef_j \times (max_j - min_j)/C_r(S_j) \tag{6}$$

The global benefit SB (RB) of processing sorted (random) accesses is defined as the sum of average benefits of the sources allowing this kind of access.

$$SB = \sum_{S_j \in \mathcal{S}_S \cup \mathcal{S}_{SR}} \overline{B_s}(S_j)$$

$$RB = \sum_{S_j \in \mathcal{S}_R} \overline{B_r}(S_j) + \sum_{S_j \in \mathcal{S}_{SR}} \overline{B_{rs}}(S_j)$$

where $\mathcal{S}_S$, $\mathcal{S}_R$ and $\mathcal{S}_{SR}$ are respectively the disjoint sets of S-sources, R-sources and SR-sources.

In conclusion, after developing the terms of SB and RB following formulas (3), (5) and (6) above, the access ratio r used by BR-Cost* becomes:

$$r = SB/RB = \frac{\sum_{S_j \in \mathcal{S}_S \cup \mathcal{S}_{SR}} \frac{A_j}{C_s(S_j)}}{\sum_{S_j \in \mathcal{S}_R} \frac{A_j}{C_r(S_j)} + \sum_{S_j \in \mathcal{S}_{SR}} \frac{A_j}{2C_r(S_j)}} \tag{7}$$

Access	Retrieved	candidates	U_{unseen}
S_1/S	$(o_2, 0.4)$	$\{(o_2, [0.4, 2.4])\}$	2.4
S_2/S	$(o_3, 0.9)$	$\{(o_2, [0.4, 2.3]), (o_3, [0.9, 2.3])\}$	2.3
S_2/S	$(o_1, 0.2)$	$\{(o_3, [0.9, 2.3]), (o_2, [0.4, 1.6]), (o_1, [0.2, 1.6])\}$	1.6
S_3/R	$(o_3, 0.8)$	$\{(o_3, [1.7, 2.1]), (o_2, [0.4, 1.6]), (o_1, [0.2, 1.6])\}$	1.6
S_1/S	$(o_1, 0.3)$	$\{(o_3, [1.7, 2]), (o_2, [0.4, 1.6]), (o_1, [0.5, 1.5])\}$	1.5
...	...	...	...

Fig. 5. First steps of BR-Cost* for $k = 2$ over the example sources in Fig. 2

where $A_j = coef_j \times (max_j - min_j)$ is the amplitude of the interval produced by S_j in the aggregated score.

Example. As an example, we present in Fig. 5 the first steps of BR-Cost* for the query in Fig. 1 over the sources in Fig. 2. We have $\mathcal{S}_S = \{S_1\}, \mathcal{S}_R = \{S_3\}, \mathcal{S}_{SR} = \{S_2\}$. The candidates set is sorted by decreasing value of the upper bound, i.e. the first k ones form $\mathcal{U}_k$. Let us consider that $k = 2$, $C_s(S_1) = C_s(S_2) = 1$ and $C_r(S_2) = C_r(S_3) = 2$. We have $\forall j$, $coef_j = 1$, $max_j = 1$, $min_j = 0$, so $A_j = 1$. The access ratio $r = SB/RB = (1/1+1/1)/(1/2+1/4) = 8/3$, so a random probe is allowed only after at least r sorted accesses, i.e. 3 sorted accesses before a random one.

- First access is sorted, because $|candidates| < k$. Benefits for S_1 and S_2 computed by **BestSortedSource** are both 0, because $N_j = 0$ (no top-k candidates yet). Remind that BR-Cost* uses the same **BestSortedSource** as BR-Cost, based on benefit $Bs_j = coef_j \times N_j \times \delta_j/C_s(S_j)$. Source S_1 is then randomly chosen.
- The second access is also sorted ($|candidates| < k$), but this time $N_2 = 1$ (o_2 not yet read in S_2), while $N_1 = 0$. Since $\delta_2 = 1 - 0.9 = 0.1$, we have $Bs_2 = 0.1/1 = 0.1$, while $Bs_1 = 0$, so S_2 is chosen.
- Now $|candidates|=k$, but *CostCondition* requires a third sorted access before a random probe become possible. We have $N_1 = 1$ (o_3 not yet retrieved in S_1), $N_2 = 1$ (o_2 not yet retrieved in S_2), $\delta_1 = 0.4 - 0.3 = 0.1$, $\delta_2 = 0.9 - 0.2 = 0.7$, so $Bs_1 = 0.1$ and $Bs_2 = 0.7$, i.e. S_2 is chosen for a sorted access.
- Since *CostCondition* allows now random probes and $U_k = U_{unseen} = 1.6$, *SortedAccessCondition* returns false and a random access is scheduled. **Choose-Candidate** returns the least refined candidate in $\mathcal{U}_k = \{o_3, o_2\}$. Both objects have been read in one source, but o_3 is returned, because it has a larger interval than o_2. Since $crtmax_2 = 0.2$ and $crtmax_3 = 1$, benefits of random sources are $Br_2 = 0.2/2 = 0.1$ and $Br_3 = 1/2 = 0.5$. Anyway, o_3 has already been read in S_2, so S_3 is the only possible choice for a random probe of o_3.
- *CostCondition* forces again at least three sorted accesses. For the benefit, we have $N_1 = 1$ (o_3 not yet retrieved in S_1), $N_2 = 1$ (o_2 not yet retrieved in S_2), $\delta_1 = 0.4 - 0.3 = 0.1$, $\delta_2 = 0.2 - 0.15 = 0.05$, so $Bs_1 = 0.1$ and $Bs_2 = 0.05$, i.e. S_1 is chosen for a sorted access.
- Execution continues in a similar way until *StopCondition* is satisfied.

3.2 Necessary Choices

As mentioned above, *Necessary Choices* (NC) [19] was the first proposal for a generic algorithm, yet constrained to the case of complete top-k scoring. In this context, NC identifies *necessary accesses* at some moment, as being those for candidates in $\mathcal{U}_k$. Algorithms in the general NC framework do only necessary accesses: each step selects an element in $\mathcal{U}_k$ with incomplete scoring and performs an access for it.

In this framework, NC proposes an algorithm SR/G that favors sorted against random accesses for each candidate. SR/G is guided by two parameters: $D = \{d_1, ..., d_m\}$, which indicates *a depth of sorted access* in each S- or SR-source, and H, which indicates *a fixed order of probes* in the random (R and SR) sources for all the candidates. The meaning of D is that sorted access to a source S_j where $crtmax_j \geq d_j$ has always priority against random probes.

Among all the possible pairs (D, H), SR/G selects the optimal one by using sampling optimization. The optimization process converges iteratively: for some initial H, one determines the optimal D, then an optimal H for this D, etc.

Despite its genericity, NC is hardly comparable with BR. In the context of incomplete top-k scoring adopted by BR, NC's analysis of necessary accesses is no longer valid. Source sampling used by SR/G is not always possible and does not guarantee similar score distribution. We propose here a variant of SR/G, adapted to the context of BR by considering incomplete scoring and a heuristic approximation of (D, H) inspired by BR-Cost*. The intention is to compare *the strategies* proposed by BR-Cost* and SR/G in a context as similar as possible.

The SR/G variant we propose is expressed in the GF framework as follows:

- Besides D and H, a local variable keeps the *best candidate*, i.e. the candidate in $\mathcal{U}_k$ with incomplete scoring having the highest upper bound. SR/G does a first sorted access to some source; the best object is initialized with this first retrieved object and updated after each iteration. Note that at least one object in $\mathcal{U}_k$ has incomplete scoring if the *StopCondition* has not been yet reached.
- *SortedAccessCondition* returns true if the set of sorted sources in which the best candidate has not been yet retrieved and where $crtmax_j \geq d_j$ is not empty.
- **BestSortedSource** returns one of the sources in this set.
- **ChooseCandidate** returns the best candidate.
- **BestRandomSource** returns the first random source not yet probed for the best candidate, following the order defined by H.
- *StopCondition*, for incomplete scoring, is given by (1).

We propose an heuristic approximation of D and H, based on the notion of source benefit used for BR-Cost*.

For H we consider the descending order of the random source benefit computed with (5) and (6).

Estimation of D is based on three hypotheses:

1. The number of sorted accesses to a source must be proportional to the source benefit given by (3).
2. Sorted accesses until depth d_j in each source should produce a decrease of the threshold enough for discriminating the top-k result, which is at least until $U_{unseen} = R_k$, where R_k is the k-th highest real score of an object.
3. If $n_j = N - Nr_j$ is the number of sorted accesses in S_j for reaching depth d_j (see Fig. 4), the relation between n_j and d_j depends on the score distribution in sources, generally unknown and approximated here with uniform distribution.

If we note $\Delta_j = max_j - d_j$ the score decrease to reach depth d_j, the three hypotheses above give:

1. $\forall j, n_j = C \times \overline{B_s(S_j)}$, where C is a constant.
2. $U_{max} - R_k = \sum coef_j \times \Delta_j$, where $U_{max} = \mathcal{F}(max_1, ..., max_m)$ is the highest possible aggregated score.
3. $\forall j, n_j / N = (max_j - d_j)/(max_j - min_j)$.

Resolving this equation system produces the following estimation for the depth:

$$d_j = max_j - \frac{A_j^2}{coef_j \times C_s(S_j)} \times \frac{U_{max} - R_k}{\sum_j A_j^2 / C_s(S_j)} \qquad (8)$$

Real score R_k may be estimated by various methods. This is not important in our experimental evaluation, since we precompute the R_k value, hence considering the best case for SR/G.

Example. We take the same example as for BR-Cost*, in the same conditions, to illustrate the first steps of the new variant of NC (see Fig. 6).

We first compute $D = \{d_1, d_2\}$ and H. For the two sources with sorted access (S_1 and S_2) we have similar parameters, $max_j = A_j = 1$, $coef_j = 1$, $C_s(S_j) = 1$, $U_{max} = 3$ and $R_k = 1.4$ (the real score of the second best object, which is o_1). Formula 8 gives $d_1 = d_2 = 1 - (3 - 1.4)/2 = 0.2$. For H, we have $\overline{B_r(S_3)} = 1/2$ and $\overline{B_{rs}(S_2)} = 1/4$, so the order given by H is $[S_3, S_2]$.

Access	Retrieved	*candidates*	U_{unseen}
S_1/S	$(o_2, 0.4)$	$\{(o_2, [0.4, 2.4])\}$	2.4
S_2/S	$(o_3, 0.9)$	$\{(o_2, [0.4, 2.3]), (o_3, [0.9, 2.3])\}$	2.3
S_2/S	$(o_1, 0.2)$	$\{(o_3, [0.9, 2.3]), (o_2, [0.4, 1.6]), (o_1, [0.2, 1.6])\}$	1.6
S_1/S	$(o_1, 0.3)$	$\{(o_3, [0.9, 2.2]), (o_2, [0.4, 1.6]), (o_1, [0.5, 1.5])\}$	1.5
S_1/S	$(o_4, 0.25)$	$\{(o_3, [0.9, 2.15]), (o_2, [0.4, 1.6]), (o_1, [0.5, 1.5]), (o_4, [0.25, 1.45])\}$	1.45
S_1/S	$(o_3, 0.2)$	$\{(o_3, [1.1, 2.1]), (o_2, [0.4, 1.6]), (o_1, [0.5, 1.5]), (o_4, [0.25, 1.45])\}$	1.4
S_3/R	$(o_3, 0.8)$	$\{(o_3, [1.9, 1.9]), (o_2, [0.4, 1.6]), (o_1, [0.5, 1.5]), (o_4, [0.25, 1.45])\}$	1.4
S_2/S	$(o_4, 0.15)$	$\{(o_3, [1.9, 1.9]), (o_2, [0.4, 1.55]), (o_1, [0.5, 1.5]), (o_4, [0.4, 1.4])\}$	1.35
...	...	...	...

Fig. 6. First steps of NC for $k = 2$ over the example sources in Fig. 2

- A first sorted access is done, S_1 is randomly chosen for that.
- o_2 is the best object and among the missing accesses for it, the sorted one in S_2 has priority, because $crtmax_2 = 1 > d_2$, so S_2 is selected for a sorted access.
- o_2 is again the best object and for the same reason as above, S_2 is selected again for a sorted access ($crtmax_2 = 0.9 > d_2$).
- Now o_3 is the best object and among the missing accesses for it, the sorted one in S_1 has priority, because $crtmax_1 = 0.4 > d_1$, so S_1 is selected for a sorted access.
- o_3 is again the best object and for the same reason as above, S_1 is selected again for a sorted access.
- o_3 is still the best object and for the same reason as above, S_1 is selected again for a sorted access.
- o_3 is still the best object, but no more sorted access for it is possible, so the last access for it is considered - the random access for o_3 is scheduled in S_3.
- o_3 being now fully evaluated, o_2 is the new best object. The sorted access for o_2 in S_2 still has priority, because $crtmax_2 = 0.2 \geq d_2$, so S_2 is selected for a sorted access.
- o_2 is again the best object, but the sorted access to S_2 has not priority anymore, because $crtmax_2 = 0.15 < d_2$. The remaining access for o_2 (random probe in S_3) is scheduled. Then execution continues in a similar way until *StopCondition* is satisfied.

3.3   Combined Algorithm

Although Combined Algorithm (CA) [7], limited to SR-sources, is not a generic algorithm, it was a first attempt towards genericity, by proposing to combine NRA and TA strategies to adapt to the case of different costs for random and sorted access.

We propose here *CA-gen*, a generic variant of CA adapted to any source types. Like for CA, if r is the ratio between the average costs of random and sorted access, each sorted (S- and SR-) source is accessed r times, before performing all the random probes for the best candidate in $\mathcal{U}_k$ with *incomplete scoring in random sources*. Note that the best candidate may be different of χ.

Unlike CA, but similar to BR, CA-gen does not produce complete scoring for the best candidate, since its score may still be unknown in some sorted sources. Like for BR, the stop condition corresponds to incomplete top-k scoring.

- The cycle of r sorted accesses in each source can be simulated in GF with local variables indicating the next source to access and the number of accesses already performed in the cycle.
- *SortedAccessCondition* returns true if the cycle is not yet finished.
- **BestSortedSource** simply returns the next source.
- **ChooseCandidate** returns the best candidate, as defined above.

Access	Retrieved	candidates	U_{unseen}
S_1/S	$(o_2, 0.4)$	$\{(o_2, [0.4, 2.4])\}$	2.4
S_1/S	$(o_1, 0.3)$	$\{(o_2, [0.4, 2.4]), (o_1, [0.3, 2.3])\}$	2.3
S_2/S	$(o_3, 0.9)$	$\{(o_2, [0.4, 2.3]), (o_1, [0.3, 2.2]), (o_3, [0.9, 2.2])\}$	2.2
S_2/S	$(o_1, 0.2)$	$\{(o_3, [0.9, 2.2]), (o_2, [0.4, 1.6]), (o_1, [0.5, 1.5])\}$	1.5
S_3/R	$(o_3, 0.8)$	$\{(o_3, [1.7, 2]), (o_2, [0.4, 1.6]), (o_1, [0.5, 1.5])\}$	1.5
...	...	...	...

Fig. 7. First steps of CA-gen for $k = 2$ over the example sources in Fig. 2

– **BestRandomSource** returns the first random source not yet probed for the best candidate. If no such source exists, the cycle stops.
– *StopCondition*, for incomplete scoring, is given by (1).

Example. We take the same example as for BR-Cost* and NC, in the same conditions, to illustrate the first steps of CA-gen (see Fig. 7).

The access ratio $r = C_r(S_j)/C_s(S_j) = 2$, so two sorted accesses are scheduled in each source before fully evaluating the best object.

– $r = 2$ sorted accesses to S_1, then to S_2 are done.
– o_3 is the best object and we schedule all the random probes for it. The only possible one is the access to S_3, because o_3 has been already read in S_2 through a previous sorted access. Note that o_3 is not fully evaluated after this step because its value in S_1 is missing and cannot be obtained by random access.
– Execution continues in the same way, by cycles of two sorted accesses in each source followed by random probes for the best object, until *StopCondition* is satisfied.

4   Approximation by Early Stopping

Top-k processing in our context is usually expensive because of predicate evaluation, therefore reducing the execution cost by accepting approximate results is a promising approach. We adopt the method proposed by TA$_\theta$ [7], based on relaxing the threshold condition in TA with a factor $\theta > 1$, i.e. the algorithm stops when the score of at least k candidates exceeds U_{unseen}/θ. This produces a θ-approximation of the top-k result, i.e. a set K_a of k objects such that $\forall x \in K_a, \forall y \notin K_a, \theta \times score(x) \geq score(y)$. As explained in the related work section, a θ-approximation guarantees that the ratio between the score of the best missed object in the approximate result (best false negative) and that of the worst false positive cannot exceed θ.

Note that this method is equivalent to an *early stopping* of the exact algorithm, i.e. TA and TA$_\theta$ have the same execution until the end of TA$_\theta$, which occurs first.

We generalize here the TA$_\theta$ approach to the case of incomplete scoring within the GF framework, i.e. to any top-k algorithm in our context, and thus enable the comparison of the behavior of top-k algorithms in the case of approximate results by early stopping.

Note that TA$_\theta$ considers that all source scores belong to the $[0, 1]$ interval. In the general case, in order to preserve the meaning of θ-approximations, we simply consider that scores in source S_j belong to a $[0, max_j]$ interval.

Consider an approximate solution K_a composed of k candidates with possibly incomplete scoring at some point during the execution of the algorithm in the GF framework. Then the condition for detecting K_a as being surely a θ-approximation of the top-k result is given by the following theorem.

Theorem 1. *An approximate solution K_a composed of k candidates with incomplete scoring during the execution of a top-k algorithm is <u>surely</u> a θ-approximation of the top-k result **iff***

$$\theta \times min_{c \in K_a}(L(c)) \geq max_{c \notin K_a}(U(c)) \tag{9}$$

Proof. At the given moment during the execution, $min_{c \in K_a}(L(c))$ represents the minimum possible score for a candidate in K_a, while $max_{c \notin K_a}(U(c))$ is the maximum possible score for an object not in K_a (including unseen objects).

We first show that if condition (9) is true, then K_a is a θ-approximation of the exact result.

For any candidate c, we have $L(c) \leq score(c) \leq U(c)$. More generally, for any unseen object o, we have $U(o) = U_{unseen}$, its maximum possible score. Then $\forall x \in K_a, score(x) \geq L(x) \geq min_{c \in K_a}(L(c))$ and $\forall y \notin K_a, max_{c \notin K_a}(U(c)) \geq U(y) \geq score(y)$. If the theorem condition holds, then $\forall x \in K_a, y \notin K_a, \theta \times score(x) \geq score(y)$, i.e. K_a is a θ-approximation.

We demonstrate the reverse implication by using proof by contradiction: if condition (9) is not true, then K_a is not surely a θ-approximation.

Consider now $x = argmin_{c \in K_a}(L(c))$ the candidate with the worst minimal score in K_a and $y = argmax_{c \notin K_a}(U(c))$ the object with the best maximal score outside of K_a. If the theorem condition does not hold for K_a, then $\theta \times L(x) < U(y)$, so it is possible that $\theta \times score(x) < score(y)$, i.e. K_a may not be a θ-approximation.

In the GF context, algorithms manage only the set of candidates discovered in sorted sources. Considering $K_a \subset candidates$, the stop condition (1) becomes:

$$\theta \times min_{c \in K_a}(L(c)) \geq max(U_{unseen}, max_{c \in candidates - K_a}(U(c))) \tag{10}$$

The difference with Theorem 1 is that here U_{unseen} gives the upper bound score for all the objects not yet discovered and thus not members of *candidates*.

Theorem 2. *Eliminating non-viable candidates does not affect the stop condition (10).*

Proof. Suppose that at some moment a non viable candidate x affects the stop condition. Since $x \notin K_a$, x can only impact the right side of the inequality and only if $U(x) > U_{unseen}$ and $U(x) > U(y), \forall y \in candidates - K_a$.

But $U(x) < L_k$ (x non-viable), so all the objects in *candidates* $- K_a$ are non-viable and $L_k > U_{unseen}$, which in accordance to (1) means that at the current moment the exact top-k has been already found, i.e. the algorithm is already stopped.

To estimate the precision of an approximate solution, we propose a distance measure based on two principles:

- Order of elements in the top-k solution is not important.
- Only wrong elements (false positives) in the approximate solution affect precision, i.e. the quality of the approximate result is given by the quality of the false positives.

Distance is measured by the difference between the real scores of wrong elements and R_k, the k-th score in the exact solution, normalized to the $[0, 1]$ interval by dividing it by R_k. Indeed, R_k is the maximum possible distance to R_k, since the lowest possible global score is 0.

The distance between an element $o \in K_a$ and the real top-k result K is defined as follows:

$$\text{dist}(o, K) = \begin{cases} (R_k - score(o))/R_k, & \text{if } o \notin K \\ 0, & \text{if } o \in K \end{cases} \tag{11}$$

The global distance between an approximate solution K_a and K is defined as the average of the individual distances between elements of K_a and K.

$$\text{dist}(K_a, K) = \frac{1}{k} \sum_{o \in K_a} \text{dist}(o, K) \tag{12}$$

We measure *the quality* of an approximate solution K_a as being $1 - dist(K_a, K)$.

The relation between our distance measure and θ-approximations is given by the following theorem.

Theorem 3. *If K_a is a θ-approximation of the real solution K, then $dist(K_a, K)$ $\leq \theta - 1$. Moreover, the $\theta - 1$ value is optimal in the general case, i.e. it is the smallest upper bound of $dist(K_a, K)$ that can be guaranteed.*

Proof. If $K_a = K$ then $dist(K_a, K) = 0$ and the inequality is true. Otherwise, considering $x \in K - K_a$, then $score(x) \geq R_k$. K_a being a θ-approximation of K, $\forall o \in K_a, \theta \times score(o) \geq score(x) \geq R_k$, so $R_k - score(o) \leq (\theta - 1)score(o)$.
In conclusion, $dist(K_a, K) = \frac{1}{k} \sum_{o \in K_a} dist(o, K) = \frac{1}{k} \sum_{o \in K_a - K} \frac{R_k - score(o)}{R_k}$
$\leq \frac{1}{k} \sum_{o \in K_a - K} \frac{(\theta-1)score(o)}{R_k} = \frac{\theta-1}{kR_k} \sum_{o \in K_a - K} score(o) \leq \frac{\theta-1}{kR_k} k R_k = \theta - 1$.

Moreover, no distance smaller than $\theta - 1$ can be guaranteed. Indeed, in the general case it is possible for K to be composed of k objects of score s, while the approximate solution K_a may be a set of k objects of score s/θ. Given the definition, this possible K_a is a θ-approximation of K, with $dist(K_a, K) = \theta - 1$.

We propose in this paper a comparative study of the approximation potential of multi-criteria top-k algorithms.

We draw cost-distance curves for these algorithms and compare their shapes. A point on the cost-distance curve indicates the quality of the approximate result on early stopping at that moment/cost. Since arbitrary early stopping comes with no guarantees on the precision of the approximate result, we also produce θ-approximations in each case and compare costs for measured and guaranteed precision.

5 Experiments

We experimentally compare the BR strategy with that of the other generic algorithms in terms of *execution cost*. Then, we compare the *approximation potential* of various categories of state-of-the-art top-k algorithms, both generic and specific.

Data Sets and Parameters. We use synthetic sources, independently generated as lists of scores in the $[0, 1]$ interval for the N objects, then organized for S, R or SR access, depending on the source type. We consider two types of score distribution in a source: uniform or exponential $(p(x) = \lambda e^{-\lambda x})$, for $\lambda = 1$ and restricted to the $[0, 1]$ interval. Exponential distribution illustrates S-sources where scores have fast decrease at the beginning, potentially more discriminant than sources with uniform distribution. The choice of synthetic data is motivated by the need for an experimental testbed with a relatively high number of criteria (up to 18 in our tests), which was not provided by the real data sets we could find.

We measure the execution costs for each algorithm as the sum of costs of all the source accesses for computing the top-k result. We consider that all the sorted (random) accesses have the same cost C_s (C_r). Each result in the experiments is the average of 10 measures over different randomly generated sources. We consider weighted sum as the aggregation function, with coefficients randomly generated for each of the 10 measures.

The following parameters are considered in the experiments:

- The number of database objects is $N = 10\,000$.
- Queries are looking for the best $k = 50$ objects.
- We consider 6 S-sources, 6 R-sources and 6 SR-sources.
- We consider the most common cost settings, with random accesses more expensive than sorted ones: $C_r = 10$, $C_s = 1$.
- Two configurations for data distribution in sources are considered: *uniform* for all the sources or *mixed*, i.e. exponential distribution for half of the sorted sources (3 S-sources and 3 SR-sources), uniform for the other sources.

Comparison of the Execution Cost. We compare the execution cost of BR-Cost* with the NC variant and CA-gen in three configurations of source types: no R-sources, no S-sources, and all the source types. We also add to the

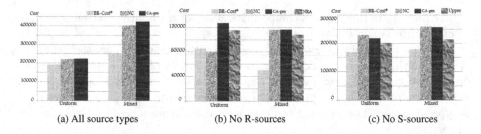

(a) All source types   (b) No R-sources   (c) No S-sources

Fig. 8. Execution cost comparison

comparison the reference non-generic algorithms compatible with that setting. In each configuration, uniform and mixed data distribution are considered.

- *All source types* (S, R and SR).
  Figure 8a shows that BR-Cost* behaves visibly better (10 %) than both NC and CA-gen for uniform distribution, while the difference becomes important for mixed distribution: approximately 37 % better than NC and 40 % better than CA-gen.

- *No R-sources* (only S and SR).
  Note that here cost and source settings are in favor of algorithms that realize only sorted access (NRA) or strongly favors them (NC). Figure 8b shows that in the uniform distribution case BR-Cost* and NC are the best, with very close costs, much better than CA-gen (around 33 %), which is even outperformed by NRA. For mixed distribution, BR-Cost* is clearly much better than NC and CA-gen (almost 60 %), which are outperformed by NRA.

- *No S-sources* (only R and SR).
  Figure 8c shows that in all the cases BR-Cost* outperforms the other algorithms and that NC and CA-gen are less adapted to this setting, performing worse than Upper. The benefit of using BR-Cost* is bigger in the mixed distribution case (around 28 % better than NC and CA-gen) compared to uniform distribution (24 %). Compared to Upper, the benefit is similar in both cases, around 15 %.

In conclusion, BR-Cost* successfully adapts to various source types and data distribution settings, and outperforms not only the other generic approaches, but also specific algorithms designed for that case. We also note a weakness for the other generic strategies in one of the studied cases: no S-sources for NC and no R-sources for CA-gen. Paradoxically, mixed distribution does not improve cost in most cases; we guess that discriminant distributions are counter-balanced here by the lack of correlation between sources and by their relatively high number.

Approximation Potential. We measure the potential of approximation by early stopping of the top-k algorithms by drawing their cost-distance curves.

Distance between approximate and real solution is computed with formulas (11)–(12). We measure this distance in several points during the algorithm's

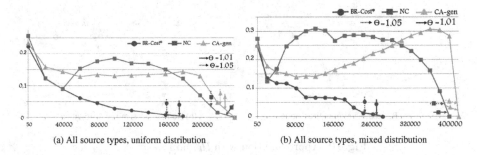

(a) All source types, uniform distribution   (b) All source types, mixed distribution

Fig. 9. Approximation with $\mathcal{U}_k$, all source types

normal execution, every 2000 cost units (or every 1000 for the no R-source case, where cost is smaller), then we extrapolate a curve relying these points. Each point on the curve represents the distance between the approximate solution and the real one if the algorithm stops at that moment. A curve "below" another one indicates a better approximation potential.

The form of the curve also indicates *approximation stability*. A monotone descending curve means stable approximation, that improves if execution continues, while non-monotony indicates an algorithm badly adapted for approximation by early stopping.

For each cost-distance curve we measure the end point that corresponds to a θ-approximation obtained with the stop condition (10). We consider two values, $\theta = 1.05$ and $\theta = 1.01$, that correspond to a guaranteed distance of 0.05, respectively 0.01 (see Theorem 3). We compare the position of these points with that of the intersection between the curve and the corresponding distance.

We consider two cases for the approximate solution. The first one is the set $\mathcal{U}_k$ of k candidates with the highest upper bound. This is a natural choice for the approximate solution, since $\mathcal{U}_k$ is the set of candidates that top-k algorithm focus on during execution. More precisely, all the algorithms proposed so far base their strategies on $\mathcal{U}_k$, either for deciding a sorted access, or for the choice of a candidate for random probes. Intuitively, candidates with high upper bounds must be "refined" because their upper bound make them potentially belong to the final top-k. The algorithm *must* decide if they really belong to the result or not - if not, the algorithm cannot end without refining the candidate's score to make it non-viable.

The second proposal for an approximate solution is the set of k candidates with the highest lower bound $\mathcal{L}_k$. Intuitively, belonging to $\mathcal{L}_k$ means that the candidate was already refined with good scores in some sources. This may be a good indication that the candidate belongs to the final top-k, better than for $\mathcal{U}_k$ where high upper bounds may be the result of little refinement, thus with high uncertainty.

Approximation with $\mathcal{U}_k$. Figures. 9, 10 and 11 present the cost-distance curves for uniform and mixed data distributions in the three cases of source types.

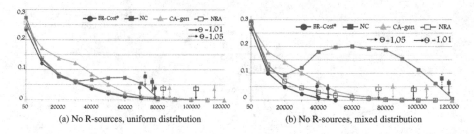

Fig. 10. Approximation with $\mathcal{U}_k$, no R-sources

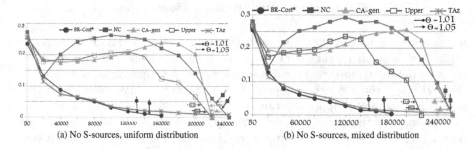

Fig. 11. Approximation with $\mathcal{U}_k$, no S-sources

Final costs for algorithms may be less visible in these figures, but they can also be retrieved in Fig. 8.

– *All source types.*

We compare the generic algorithms BR-Cost*, the NC variant and CA-gen. For uniform distribution (Fig. 9a), BR-Cost* approximation distance quickly decreases and the algorithm has clearly better approximation properties than CA-gen (much higher distance, only decreasing at the end) or NC (totally unstable).

Mixed distribution (Fig. 9b) accentuates the problems of NC and CA-gen (which becomes unstable), while BR-Cost* keeps a good curve shape. However, θ-approximation significantly reduces the cost saving for BR-Cost*, e.g. for $\theta=1.05$ algorithm stops at cost 160 000, while the corresponding distance of 0.05 is already reached at cost 70 000.

– *No R-sources* (only S and SR).

Besides the three generic algorithms, we also consider here the NRA algorithm. Excepting NC, algorithms produce in this case stable approximations. For uniform distribution (Fig. 10a), BR-Cost* and NRA have very close curves, i.e. close approximation potential, but BR-Cost* produces θ-approximations with better costs. Similarly, CA-gen has good approximation potential, especially in the second half of the execution, but θ-approximations are more expensive than for NRA. NC is more stable than in the previous case and and its low execution time helps it producing less expensive $\theta-$approximations.

For mixed distribution (Fig. 10b), BR-Cost* improves its potential compared to NRA, while NC becomes highly unstable.

- *No S-sources* (only R and SR).
Besides the three generic algorithms, we also study here the Upper and TAz algorithms. Despite the fact that it is much more expensive (around six times the cost of BR-Cost*), TAz is considered because of the good approximation potential of algorithms with many SR-sources, which reduce as much as possible the uncertainty of the candidates' scores.

For both uniform (Fig. 11a) and mixed distribution (Fig. 11b), behavior is very similar. CA-gen, NC and Upper are highly unstable, while BR-Cost* and TAz have very close curves, with very good approximation potential. However, because of its high execution cost, θ-approximations of TAz are much more expensive than for BR-Cost* (because of their high values, final cost and θ-approximations for TAz are not visible in the figure).

In conclusion, BR-Cost* has clearly the best approximation potential with $\mathcal{U}_k$ among the generic algorithms, with good properties for the different data distributions. The other generic algorithms are badly adapted to approximation with $\mathcal{U}_k$: NC and CA-gen are systematically unstable.

We guess that the good approximation properties of BR-Cost* come from its breadth-first strategy. Handling the current top-k $\mathcal{U}_k$ as a whole, instead of focusing on the best candidate only, produces a more stable evolution of $\mathcal{U}_k$ toward the final solution.

The price to pay for guaranteed precision in θ-approximations is important for algorithms with good approximation curves - we notice a significant difference with the potential cost for the same approximation quality. The cost of θ-approximations appears to be dependent on the total cost of the algorithm: for algorithms with very close cost-distance curves, higher total costs systematically lead to higher θ-approximation costs.

Approximation with $\mathcal{L}_k$. Figures. 12, 13 and 14 present the cost-distance curves for the approximation with the best k lower bound scores $\mathcal{L}_k$, in the three cases of source types. The sub-figure for each case presents, besides the curves, a zoom on the final part of the execution, where the curves are very close.

- *All source types.*
For both uniform (Fig. 12a) and mixed distribution (Fig. 12b), the curves for all the algorithms are very close, stable, with good approximation potential. BR-Cost* and CA-gen have slightly better curves than NC, the difference being visible in the mixed distribution case and on the final part of the uniform case.

Comparison of θ-approximations follows the conclusion of the previous point, algorithms with better execution costs produce better θ-approximations, i.e. BR-Cost* is the best, while CA-gen and NC are very close.

We notice that cost-distance curves with $\mathcal{L}_k$ are better than those with $\mathcal{U}_k$ in all the cases. This also leads to an increased difference between the cost with θ-approximation and the potential one.

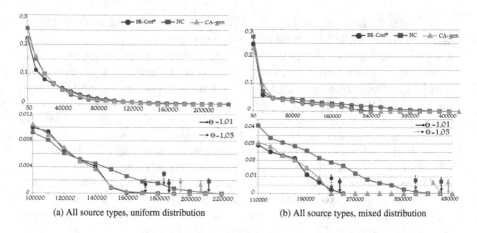

(a) All source types, uniform distribution   (b) All source types, mixed distribution

Fig. 12. Approximation with $\mathcal{L}_k$, all source types

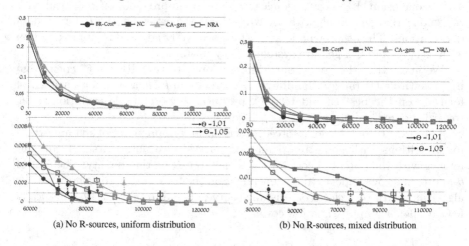

(a) No R-sources, uniform distribution   (b) No R-sources, mixed distribution

Fig. 13. Approximation with $\mathcal{L}_k$, no R-sources

– *No R-sources.*

Conclusions are similar to the all source types case for both curve shapes and θ-approximations, but differences between algorithms are more important here.

For uniform distribution (Fig. 13a), BR-Cost* has globally the best shape, followed very closely by NC and NRA, while CA-gen is slightly, but visibly worse.

For mixed distribution (Fig. 13b), the superiority of BR-Cost* is clearer, the other algorithms being close and having sections on which they have better approximation potential than the others.

– *No S-sources.*

We find similar conclusions in this case too, with the remark that generic algorithms have globally better curve shapes than Upper and TAz.

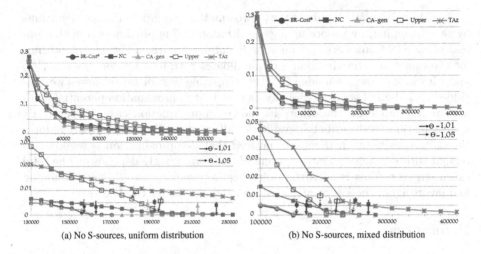

(a) No S-sources, uniform distribution   (b) No S-sources, mixed distribution

Fig. 14. Approximation with $\mathcal{L}_k$, no S-sources

For uniform distribution (Fig. 14a), BR-Cost*, CA-gen and NC are very close, with CA-gen slightly better on the middle part and NC slightly worse on the final part. Upper has globally the least favorable approximation potential, TAz becoming worse at the end only because of its higher cost.

For mixed distribution (Fig. 14b), BR-Cost* and CA-gen have clearly the best potential, followed by NC. Unlike for uniform distribution, Upper is here globally better than TAz.

In conclusion, we notice that approximation with $\mathcal{L}_k$ has better quality than with $\mathcal{U}_k$ for all the algorithms. Compared with the $\mathcal{U}_k$ case, approximation is always stable with $\mathcal{L}_k$ and has better precision at the same execution cost. Even if BR-Cost* has globally the best properties, the approximation potential of generic algorithms is very close in this case.

However, θ-approximations are not improved by $\mathcal{L}_k$ and lead to an increased difference between the potential cost and that for guaranteed precision.

6   Conclusion

In this paper we proposed a generic framework GF for top-k processing over expensive ranking predicates, able to express any top-k algorithm. We compared within this framework our generic algorithm BR with generic variants that we proposed for algorithms NC and CA, adapted to a similar context. Comparison of the algorithm strategies within GF was completed with experimental measures indicating that the breadth-first strategy of BR adapts itself very well to various source type configurations and data distributions, leading to better execution cost than the other generic or specific strategies.

We also presented a study of the approximation potential of top-k algorithms by early stopping, by proposing a generalization of θ-approximation in the context of the GF framework and an experimental comparison between algorithms for two common approximation sets: candidates with best k upper bounds ($\mathcal{U}_k$) and with best k lower bounds ($\mathcal{L}_k$). By comparing cost-distance curves we concluded that the BR strategy globally has the best approximation potential, with a clear advantage on the others in the $\mathcal{U}_k$ approximation case. However, $\mathcal{L}_k$ approximation produces better results for all the algorithms and greatly reduces the differences between them. We noticed that θ-approximation is weakly correlated with the approximation potential and significantly depends on the total execution cost. This cancels the difference between $\mathcal{U}_k$ and $\mathcal{L}_k$ approximation and favors again the BR strategy, which produces better total costs.

References

1. Akbarinia, R., Pacitti, E., Valduriez, P.: Best position algorithms for top-k queries. In: VLDB, pp. 495–506 (2007)
2. Badr, M., Vodislav, D.: A general top-k algorithm for web data sources. In: DEXA, pp. 379–393 (2011)
3. Böhm, C., Berchtold, S., Keim, D.A.: Searching in high-dimensional spaces: Index structures for improving the performance of multimedia databases. ACM Comput. Surv. **33**(3), 322–373 (2001)
4. Bruno, N., Gravano, L., Marian, A.: Evaluating top-k queries over web-accessible databases. In: ICDE, pp. 369–378 (2002)
5. Cao, P., Wang, Z.: Efficient top-k query calculation in distributed networks. In: PODC, pp. 206–215 (2004)
6. Chang, K.C.-C., won Hwang, S.: Minimal probing: supporting expensive predicates for top-k queries. In: SIGMOD Conference, pp. 346–357 (2002)
7. Fagin, R., Lotem, A., Naor, M.: Optimal aggregation algorithms for middleware. J. Comput. Syst. Sci. **66**(4), 614–656 (2003)
8. Güntzer, U., Balke, W.-T., Kießling, W.: Optimizing multi-feature queries for image databases. In: VLDB, pp. 419–428 (2000)
9. Güntzer, U., Balke, W.-T., Kießlingm, W.: Towards efficient multi-feature queries in heterogeneous environments. In: ITCC, pp. 622–628 (2001)
10. Ilyas, I.F., Aref, W.G., Elmagarmid, A.K.: Supporting top-k join queries in relational databases. VLDB J. **13**(3), 207–221 (2004)
11. Ilyas, I.F., Beskales, G., Soliman, M.A.: A survey of top- k query processing techniques in relational database systems. ACM Comput. Surv. **40**(4), 11:1–11:58 (2008)
12. Li, C., Chang, K.C.-C., Ilyas, I.F.: Supporting ad-hoc ranking aggregates. In: SIGMOD Conference, pp. 61–72 (2006)
13. Li, C., Chang, K.C.-C., Ilyas, I.F., Song, S.: Ranksql: Query algebra and optimization for relational top-k queries. In: SIGMOD Conference, pp. 131–142 (2005)
14. Mamoulis, N., Cheng, K.H., Yiu, M.L., Cheung, D.W.: Efficient aggregation of ranked inputs. In: ICDE, p. 72 (2006)
15. Marian, A., Bruno, N., Gravano, L.: Evaluating top- k queries over web-accessible databases. ACM Trans. Database Syst. **29**(2), 319–362 (2004)
16. Michel, S., Triantafillou, P., Weikum, G.: Klee: A framework for distributed top-k query algorithms. In: VLDB, pp. 637–648 (2005)

17. Natsev, A., Chang, Y.-C., Smith, J.R., Li, C.-S., Vitter, J.S.: Supporting incremental join queries on ranked inputs. In: VLDB, pp. 281–290 (2001)
18. Theobald, M., Weikum, G., Schenkel, R.: Top-k query evaluation with probabilistic guarantees. In: VLDB, pp. 648–659 (2004)
19. won Hwang, S., Chang, K.C.-C.: Optimizing top-k queries for middleware access: A unified cost-based approach. ACM Trans. Database Syst. 32(1), 5 (2007)

Continuous Predictive Line Queries for On-the-Go Traffic Estimation

Lasanthi Heendaliya$^{(\boxtimes)}$, Dan Lin, and Ali Hurson

Department of Computer Science,
Missouri University of Science and Technology,
Rolla, MO, USA
{lnhmwc,lindan,hurson}@mst.edu

Abstract. Traffic condition is one vital piece of information that any commuter would wish to obtain to plan an efficient route. However, most existing works monitor and report only current traffic, which makes it too late for commuters to change their routes when they realize they are already stuck in the traffic. Therefore, in this paper, we propose a traffic prediction approach by defining and solving a novel continuous predictive line query. The continuous predictive line query aims to accurately estimate traffic conditions in the near future based on current movement of vehicles on the roads, and continuously update the predicted traffic conditions as vehicles move. The predicted traffic condition will not only help redirect commuters in advance but also help relieve the overall traffic congestion problem. We have proposed three algorithms to answer the query and carried out both theoretical and empirical study. Our experimental results demonstrate the effectiveness and efficiency of our approach.

1 Introduction

Mobile devices are becoming more and more popular in our daily life. As of 2012, the mobile connected device usage was about 6.8 billion [15], which is numerically about 90 % of the world population. The prevalence of the mobile technology has enabled a variety of location-based services that help greatly enhance driving experience. For example, finding an optimal route and checking the real-time traffic condition are now common practice for many drivers. In this work, we aim to further advance the existing technology on traffic monitoring and incorporate the spirit of ubiquitous computing to provide even better experience for users.

In particular, we observe that most existing traffic monitoring applications only provide *current traffic condition*. However, the route calculation based on *current traffic condition* may not be optimal. Consider the following example. Bob plans to travel from Rolla to St. Louis which is about 100 miles (i.e., about 2-hour driving). When he sets off, there is a traffic jam on his way to St. Louis. If the navigation system computes the travel route for Bob based on current

© Springer-Verlag Berlin Heidelberg 2015
A. Hameurlain et al. (Eds.): TLDKS XVIII, LNCS 8980, pp. 80–114, 2015.
DOI: 10.1007/978-3-662-46485-4_4

traffic condition, the route will probably include a detour to bypass the traffic jam. However, an hour later when Bob is already on his detour route, the traffic jam has been cleared up. Bob actually needs not take the detour if the navigation system is able to calculate the route with predicted traffic condition. This kind of scenarios inspire us to design a traffic prediction system that can provide better insight in travel planning. Moreover, the traffic prediction should be proactive/pervasive in that once the user initiates a traffic condition prediction query, the system continuously monitors the prediction results and reports any changes that may be caused by the dynamic traffic. Figure 1 illustrates an example of continuous traffic prediction. Specifically, Figs. 1(a) and (b) show snapshots of three vehicles at time t_1 and t_2 respectively. The query road segment is $\overline{AB}$, and the current travel plans of the vehicles are highlighted by bold lines. As shown in Fig. 1(a), three vehicles V_1, V_2, and V_3 may enter the querying road $\overline{AB}$. However, as time passes, vehicle V_1 changes its travel plan by making a right turn earlier at time t_2. As a result, only two vehicles (V_1 and V_3) may enter the query road which requires an update of the previous query results.

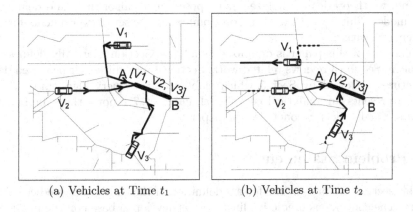

(a) Vehicles at Time t_1 (b) Vehicles at Time t_2

Fig. 1. Dynamic nature of continuous traffic prediction information

To build the above envisioned system, none of the existing approaches can be directly adopted. The closely related work that can provide traffic information includes range queries and density queries. A range query reports traffic information in a given circular or rectangular area [2,9,25], which contains traffic information on irrelevant roads rather than just the routes that the query issuer may pass by. The density query [11,14,20,26] outputs even coarse information which are regions with vehicles more than certain threshold. Moreover, most of the solutions to these query types assume an environment that objects move freely, which is not the case when the road network constraints are employed. Very few works [5,12] can be found that consider road network constraints. Those few, however, only support queries on current traffic condition but not traffic prediction.

In this paper, we propose a solution to the construction of the continuous traffic prediction system. We formalize the problem as a new type query, namely *continuous predictive line query*. The continuous predictive line query allows a user to specify a road that he/she would like to know about the traffic condition of. Then, the query returns predicted traffic condition of the querying road at the estimated time that the user may pass by. If there is any significant change of the prediction results on the querying road due to location updates of other vehicles, the updated query result will be automatically sent back to the user. To speed up the query processing and reduce the query maintenance cost, we design a novel data structure, the TPR^Q-tree, which indexes queries and efficiently handles the query result updates that evolve with time. We also propose three query algorithms that leverage the TPR^Q-tree and achieve increasing efficiency. We have carried out both theoretical and empirical study. Our experimental results demonstrate the effectiveness and efficiency of our approach.

A preliminary version of this work was published in [8]. In this paper, our new major contributions are summarized as follows. First, we propose two new query algorithms which achieve significant improvement over our previous work. Second, we theoretically analyze our proposed query algorithms and define the cost model. Third, we conduct a more comprehensive set of experiments for the system evaluation.

The rest of this paper is organized as follows. Section 2 formally defines the problem. Section 3 reviews related work. Section 4 introduces our proposed data structures, followed by Sect. 5 which elaborates the query algorithms. Then, Sect. 6 presents an analytical cost model and Sect. 7 reports the experimental results. Finally, Sect. 8 concludes the paper.

2   Problem Statement

In this section, we present the formal definition of the continuous predictive line query. The continuous predictive line query is developed based on the predictive line query as introduced in [7].

Definition 1. *[Predictive Line Query] A predictive line query $PLQ = (e_q, t_q, t_c)$ retrieves all moving objects which will be on the query road segment e_q at the query time t_q, where $t_q > t_c$ and t_c is the current time at which the query is issued.*

The predictive line query is a one-time snapshot query. It does not consider possible changes of the predicted traffic condition when the query issuer moves closer to the querying road. In order to provide timely and up-to-date information to the query issuer, we model moving objects as a linear function of time as that in many prior works [3, 10, 23, 24, 29]. Vehicles are assumed to report their locations and velocities to the server whenever there is a significant change of their moving functions. Accordingly, the continuous version of the predictive line query is defined as follows:

Definition 2. *[Continuous Predictive Line Query] A continuous predictive line query $CPL = (e_q, t_q, t_c, \rho)$ continuously monitors the moving objects which will be on the query road segment e_q at the query time t_q, and returns query results whenever the number of query results differ more than a threshold ρ. Specifically, let R_i denote the query results at time t_i ($t_c \leq t_i \leq t_q$), CPL returns the query results in the form of $\{(R_1, t_c), (R_2, t_2), ..., (R_k, t_q)\}$, and $|R_{i+1}| - |R_i| > \rho$.*

For example, a CPL query like $CPLQ = $ (AB, 8am, 7:30am, 20) means that the query issuer issued a query at 7:30am (i.e., t_c) and is interested in the traffic at road AB at 8am (i.e., t_q). The query issuer expects the server to report the change of prediction results if the difference of the number of vehicles on the querying road is more than 20. Note that it is not necessary for the query issuer to specify the threshold parameter. Instead the threshold can be automatically chosen by the server according to the past experience to reflect significant traffic change. Moreover, the server can also provide the query issuer the traffic information in an easy-to-understand form like "may have traffic jam"or "traffic flow will be good" based on the raw number of query results and the number of lanes on the specific road.

3   Related Work

Our proposed continuous predictive line query is a new type of spatial-temporal query in moving objects databases. In what follows, we review the existing common types of spatial-temporal queries and discuss their differences from the continuous predictive line query.

Spatial-temporal queries on moving objects can be classified into two main categories: snapshot queries and continuous queries. Snapshot queries execute the user-issued query only once and report the current or predicted positions of moving objects. Examples of these types of queries include range queries [2,9,25], density queries [11,14,20,26], k-nearest neighbor queries [18], reverse k-nearest neighbor (KNN) queries [27], and detour queries [21]. Specifically, a range query [2,9,25] retrieves moving objects located in a query region at a specified query time which could be either a current timestamp or a future timestamp. Applying this query to the traffic prediction in our problem, it however may include unnecessary information of vehicles on other roads that the query issuer will not pass by. Moreover, most of existing works on range queries do not consider the road-network constraints as we did in our work. A density query [11,14,20,26] only requires the user to specify a time of interest but not necessary a query region. Then, the density query is able to identify regions with the density of moving objects being extremely high (above some threshold). We can see that the density query is also not suitable for predicting traffic on a specified road segment on a user's travel route. The KNN query [27] locates k nearest moving objects to a query location, which means the number of objects in the query results is always k and hence could not provide the insight on how many vehicles on the whole road. The detour queries [21] aim to find a new travel path based on the current traffic condition which deals with an orthogonal problem compared to our work.

Our proposed continuous predictive line query is more related to continuous spatial-temporal queries which keep updating the query results as moving objects move around. For most snapshot queries, there is a corresponding continuous query version for it, such as continuous range queries and continuous KNN queries. A common approach adopted by existing continues queries is to define safe regions for moving objects [2,9,16,17,22]. When objects are moving within the safe regions, the query results remain the same. When objects move outside the safe regions, they check if there is a need to update the query results. To improve scalability, some works [6,19,28] propose distributed approaches which distribute the maintenance tasks of the query results to moving objects. Compared to our work, most of existing works on continuous queries assume that moving objects in Euclidean space rather than on the road networks, which will affect the accuracy of the traffic prediction as reported in [8]. For those few [4,13] that consider the road network constraints, they only query on current positions of moving objects but do not provide prediction of future positions. Therefore, our work will fill in the gap and advance the state-of-the-art in the traffic prediction.

In addition, our proposed TPR^Q-tree is inspired by the TPR-tree and its variants [23,24]. The TPR-tree indexes the movement functions of moving objects in Euclidean space. The MBRs in the TPR-tree are always growing as time evolves. Unlike the TPR-tree, our TPR^Q-tree indexes the continuous queries rather than the moving objects. The MBRs in the TPR^Q-tree are always shrinking as time evolves. In all, the TPR^Q-tree serves a totally different purpose compared to the TPR-tree, and the TPR^Q-tree is associated with a new suite of query algorithms.

4   Data Structures

In order to efficiently answer the continuous predictive line query, we employ the R^D-tree [7] to index the road networks and moving objects; and we propose a new data structure, namely Time-Parameterized R*-tree for Query (TPR^Q-tree), to index continuous queries issued by users. In what follows, we describe the two data structures in detail.

4.1   The R^D-tree

We adopt the R^D-tree because it supports snapshot predictive line queries on moving objects under the road network constraints. The R^D-tree is composed of an R*-tree and a set of hash tables. Figure 2 illustrates the overall structure of the R^D-tree. The road-network information is indexed by the R*-tree. Each entry in the non-leaf node is in the form of ($node\_MBR, child\_ptr$), where $node\_MBR$ is the minimum bounding rectangle (MBR) covering the MBRs of all entries in its children pointed by $child\_ptr$. Leaf nodes in R*-tree pointing to hash tables represent vehicles at each road segment. Each entry in the leaf node is in the form of $(edge\_MBR, obj\_ptr)$, where $edge\_MBR$ is the MBR of a road segment and obj_ptr links to a hash table storing objects moving on this edge.

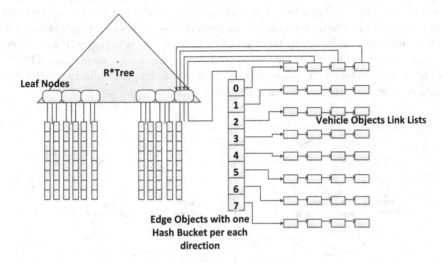

Fig. 2. R^D-tree indexing structure

The road network is represented as a graph $G(E, V)$; where E is the set of edges, and V is the set of vertices. Each edge $e \in E$ represents a road segment in the network. Here, $e = \{v_1, v_2\}$, where $v_1, v_2 \in V$; v_1 and v_2 are starting and end nodes of the road segment, respectively. Furthermore, each edge is associated with two parameters: l and s, where l is the length of the edge and s is the maximum possible speed on that edge. Each edge also maintains a list of vehicles moving on it.

A moving object (or a vehicle[1]) O is represented by the tuple $\{vId, x_1, y_1, e_c, e_n, speed, e_d, t\}$, where vId is the unique ID of the vehicle, x_1 and y_1 are the coordinates of the vehicle at the latest update timestamp t, e_c is the current road segment that the vehicle is on, e_n is the next road segment that the vehicle is heading to, and e_d is the vehicle's traveling destination. Here, it is assumed that most moving objects are willing to disclose their tentative traveling destinations to the service provider (server) in order to obtain high-quality services. However, the destination may change during the trip. Moving objects are grouped according to the geographical direction formed according to individual's destination with respect to the current position. Details of the R^D-tree can be found in [7].

4.2   The TPR^Q-tree

The continuous predictive line (CPL) queries require to continuously monitor the moving objects on the query road segment at a near future timestamp.

---

[1] Both the terms Moving Object and Vehicle will be used interchangeably.

A naive approach for answering a continuous query is to reprocess the same query every timestamp till the expiration of the life time of the continuous query. This may involve lots of unnecessary efforts if there is no change of the query results at consecutive timestamps. Observing more closely, there is a need to update the query results only when an object in the current result becomes invalid or a new object joins the result due to the change of the moving function. Given a large amount of moving objects updates occurred every timestamp, we propose TPR^Q-tree to be used to facilitate quick identification of which update affects which CPL query in order to achieve efficient query performance.

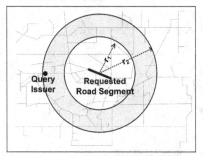 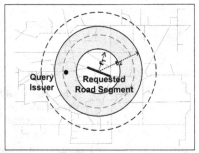

(a) Influence region at query issuing time: t_c

(b) Influence region at a later time stamp: t'_c

Fig. 3. Shrinking influence region at time t_c and $t'_c (> t_c)$

The TPR^Q-tree does not simply index the query road segment of a CPL query. Instead, the TPR^Q-tree indexes an *influence region* for each CPL query. The *influence region (IR)* is the region which covers majority of moving objects that may enter the desired query road segment at the future query time. In other words, if objects in the IR update their movement functions, the query results may be affected. To have a better understanding of the IR, let us consider a query Q that aims to predict moving objects entering the highlighted road in 30 min from now. Figure 3(a) shows the IR of Q at query issuing time. From the figure, we can see that the IR has a ring shape. Its inner radius is the road distance that can be traveled in 30 min (the query interval) by an object with the minimum speed[2]. On the contrary, its outer radius is the road distance that can be traveled by an object with fastest moving speed in 30 min. All moving objects covered by this ring have the possibility to enter the query road segment. The interesting observation here is that as time evolves, the IR will shrink as shown in Fig. 3(b). This is because the time to travel to the query road segment is shortened as the time getting closer to the future query time. More specifically, at the query issuing time, the CPL query considers objects which travel 30 min

---

[2] Note that the minimum speed is always greater than 0 since we exclude outlier objects with speed equal to 0 (e.g., an object stopped at a gas station).

to enter the road segments. After 10 min of the query issuing time, the CPL query considers objects which take 20 min to enter the road segments.

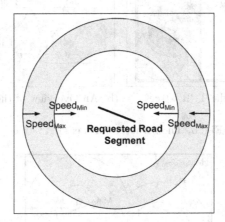

Fig. 4. Shrinking speeds of the influence region

To model the shrinking IR, it is stored as the parameterized ring which has moving speed attached to both inner and outer radius. The inner radius is associated with a minimum moving speed towards the query road segment while the outer radius is associated with a maximum speed towards the query road segments as shown by the arrows in Fig. 4. The time-parameterized IR is formally defined as follows.

Definition 3. *[Influence Region] Let $Q = (e_q, t_q, t_c, \rho)$ be a CPL query. The influence region is a time-parameterized ring in the form of $IR = (c, r_1, speed_{min}, r_2, speed_{max})$, where c is the middle point of the querying road e_q, r_1 and r_2 are the radius of the inner and outer circles respectively, and $speed_{min}$ and $speed_{max}$ are the shrinking speed of the inner and outer circles respectively. The radii are computed as follows: $r_1 = RoadDist(speed_{min} \cdot (t_q - t_c))$ and $r_2 = RoadDist(speed_{max} \cdot (t_q - t_c))$.*

Data Structure of the TPR^Q-tree. Figure 5 illustrates the structure of a TPR^Q-tree. The base structure of the TPR^Q-tree is the R^*-tree. There are three types of nodes in the TPR^Q-tree: leaf nodes; immediate parent node of the leaf nodes; higher-level internal nodes. We elaborate their structures as follows:

- Leaf level: An entry in the leaf node of the TPR^Q-tree stores information of a group of CPL queries. The information includes each query's parameters (e_q, t_q, t_c, ρ), the corresponding influence region IR, a list of query issuers, and a pointer to the query results.
- Second level: Each entry in the parent node of the leaf nodes stores a pointer to the leaf node and a time-parameterized minimum bounding rectangle (MBR)

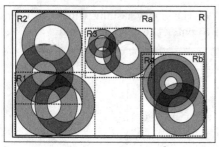
(a) Leaf node in the TPR^Q-tree

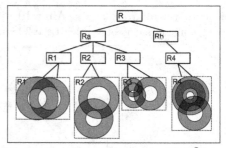
(b) An Overview of the Entire TPR^Q-tree

Fig. 5. The Structure of the TPR^Q-tree

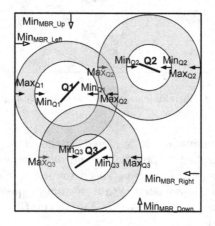

Fig. 6. Shrinking Speeds of the MBR

that bounds all the IRs of the queries in the leaf node. The time-parameterized MBR has a speed attached to each edge as shown in Fig. 6. The speed of each edge is the minimum speed among the speeds of outer rings of all IRs in the MBR. The moving direction of each edge is pointing to the center of the MBR so that the MBR shrinks as time passes and bounds the shrinking IRs. The time-parameterized MBR is stored as a six-tuple $(x_1, y_1, x_2, y_2, speed, t_u)$ where (x_1, y_1) is the coordinates of the left lower corner of the MBR, (x_2, y_2) is the right upper corner of the MBR, $speed$ is the speed of each edge and t_u is the latest time that the parameters of the MBR is updated.

– Higher levels: An entry in higher level internal nodes contains a pointer to the child node and a time-parameterized MBR that bounds MBRs in the child node. Each edge of the MBR is associated with a minimum speed among the speeds of its child MBRs and each edge is moving towards the center as well.

Construction and Maintenance of the TPR^Q-Tree. There are three types of basic operations in the TPR^Q-tree: inserting a new query, deleting an existing query, and updating an existing query.

An outline of the insertion algorithm is shown in Fig. 7. Given a new CPL query, its IR (denoted as IR_{new}) is calculated (line 1). The TPR^Q-tree is then searched to find the proper leaf node to store the new query. The algorithm to identifying the leaf node (chooseSubTree() on line 2) will be described shortly in the next paragraph. At the end of the search, if the same query is found in a leaf node (since other users may have already issued the same query), the results stored with the query will be directly return to the user and the ID of this user will be appended to the list of the query issuers. Note that two queries are considered the same if they are querying traffic of the same road segment at the same near future timestamp. Moreover, without affecting the service quality much, the query cost can be significantly saved by requiring users to specify the query time at a lower resolution of the time (e.g., every 10 min instead of every second) so that the probability of having same queries at same timestamp will be increased. If the new query does not exist in the tree, it will be inserted to the identified leaf node (line 5–10) and the predictive line query algorithm will be executed to obtain the initial query results for this new query (line 11). It is worth noting that the insertion at the leaf node may trigger updates to its parent nodes all the way up to the root node in that the speed and the size of the MBRs of its ancestor nodes may need to be adjusted to ensure the newly inserted query is enclosed. In addition, if the insertion encounters a node that is full (line 10), the node will be split. The node splitting algorithm is similar to that in the R*-tree. The only difference is that the speeds of the MBRs after the splitting need to be re-calculated.

Procedure TPR^Q-tree Insert
Input : Q
Input : Q_{Result}

1. $IR_{new} \leftarrow Q.getIR(Q.time())$
2. $node \leftarrow TPRQ.chooseSubTree(TPRQ.root, IR_{new}, Q.time())$
3. **if** $node = DATA$ **then**
4. $Q_{Result} \leftarrow node.result()$
5. **else**
6. $numOfChildren \leftarrow node.numChildren()$
7.   **if** $numOfChildren < QueryRTree.MAX$ **then**
8. $node.addAChild(Q)$
9.   **else**
10. $NodeSplit(node, Q, Q.time)$
11. $Q_{Result} \leftarrow RD.snapshotQuery(Q)$
12. **return** Q_{Result}

Fig. 7. Description of the TPR^Q -tree Insert operation

We now proceed to elaborate the chooseSubTree() algorithm (Fig. 8). This process starts from the root. For each node being examined during the search, the

---

Procedure chooseSubTree
Input : *parent*, IR_{new}, *time*
Output : *atreenode*

1. $minAreaEnl \leftarrow BIGNUMBER$
2. $minArea \leftarrow BIGNUMBER$
3.   **for each** *children* $\in$ *parent* **do**
4. $areaEnl \leftarrow findAreaEnlargement(children, IR_{new}, time)$
5. $area \leftarrow findArea(children, IR_{new}, time)$
6.   **if** $(minAreaEnl > areaEnl)$ **or** $(minAreaEnl = areaEnl$ **and** $(minArea > area))$ **then**
7. $newNode \leftarrow children$
8. $minAreaEnl \leftarrow areaEnl$
9. $minArea \leftarrow area$
10.   **if** *newNode* **not** *LEAF* **then**
11.   **return** $chooseSubTree(newNode, IR_{new}, time)$
12.   **else if** *newNode* = *LEAF* **then**
13. $duplicate \leftarrow newNode.findDuplicate(IR_{new}, time)$
14.   **if** *duplicate* = *nil* **then return** *children*
15.   **return** *duplicate*

Fig. 8. Description of the `chooseSubTree` operation

chooseSubTree() algorithm first computes the MBR of each entry of this node at the current timestamp based on the shrinking speed of the corresponding MBR. The entries with the MBRs that fully cover IR_{new} will be considered first (i.e., $areaEnl = 0$). If none of the MBRs fully cover IR_{new}, the entry with the MBR that needs the minimum enlargement to include IR_{new} will be considered. If there are several candidate entries, the entry with the MBR of the minimum area will be chosen to break the tie. At the end of the search, the algorithm returns either an existing query (if the same query is found) or a leaf node for inserting the new query.

Next, we introduce the deletion algorithm. A CPL query needs to be deleted from the TPR^Q-tree either when the query issuer passed the querying road segment or when the issuer withdrew the query before the query expires. Figure 9 outlines the deletion process. Specifically, given a query to be deleted, the first step is to locate this query. The search starts from the root of the TPR^Q-tree. At each level, the entries with the MBRs that fully cover the query's IR will be considered (line 1) and their children nodes will be checked in the same way until the leaf nodes are reached. Then, check each located leaf node to identify the one that contains the query to be deleted. After deleting the query from the leaf node, the MBR of the leaf node may need to be re-calculated, and the update may propagate to the ancestor nodes of this leaf node all the way to the root of the tree. In addition, if the deletion causes a node underflow (containing entries fewer than half of the capacity), the under-flow treatment will be applied (line 6). The under-flow treatment considers the merging with the sibling node first. If the merging can not be done due to the relatively full occupation of the sibling nodes, entries of the under-flowed node will be deleted and reinserted to the tree.

Procedure TPR^Q-tree Delete
Input : Q

1. $parentNode \leftarrow TPRQ.search(Q)$
2.   **if** $parent \neq null$ **then**
3. $parentNode.remove(Q)$
4. $updateMBR()$
5.   **if** $parentNode.numChildren() < QueryRTree.MIN$ **then**
6. $underflowTreat(parentNode)$

Fig. 9. Description of the TPR^Q -tree Delete operation

A query update is processed as follows. First, we search the TPR^Q-tree to locate the leaf node containing the query. If the query with the new parameters is still covered by the MBR of the leaf node, we will update the query parameters as well as the speed of the MBR of this leaf node if the speed needs to be changed to the new query parameters. If the new query can no longer be included in the current leaf node, we delete the query and treat it as a new query to be inserted to the tree.

5 Continuous Predictive Line Query Algorithms

In this section, we present the CPL query algorithm which consists of two phases: the initial phase and the maintenance phase. The initial phase computes the query result that is valid at the query issuing time. The maintenance phase maintains the query results as time passes.

5.1 Initial Phase

Upon receiving a new query from a user u, the TPR^Q-tree will be updated as discussed in Sect. 4.2. Recall that if the new query coincides with a previously stored query in the TPR^Q-tree, there is no need to execute this query again. Instead, the stored query results will be directly returned to the user u, and hence repeated query execution is avoided. This is one of the advantages of the TPR^Q-tree. In practice, it can be expected that many people might be interested in some particular road segments. That could be because the road segments often have traffic congestion issues, or they are the hubs for many popular destinations. Thus, in this kind of situation, using the TPR^Q-tree to group the same users with respect to the same query helps save the query cost.

In other cases when the new query cannot be found in the TPR^Q-tree, we will first insert the new query to the tree, and then execute a snapshot predictive line query [7] to identify those moving objects that may enter the query road segment at the query time based on their current movement functions.

These initial query results will be reported to the user and stored along with the new query in the TPR^Q-tree. Due to the characteristics of mobile objects, the initial query results will need to be revised during the subsequent maintenance phase until the query expires.

5.2 Maintenance Phase

The query results computed at the initial phase may need to be updated upon changes of some vehicles' travel plans just as shown in Fig. 1 in the introduction.

If a vehicle changes its moving direction or speed dramatically, the vehicle will send an update to the server. Upon receiving the update message, the server performs two tasks. The first task is to update the object in the R^D-tree [7]. The second task is to check if the update affects existing queries by answering the following two questions: (1) Is this object currently included in any existing query result? (2) Is this object going to be in some queries results' after the update? Given an object update and one query, there are four cases for the above two questions:

1. The object is included in the query result, and is still the query result after the update.
2. The object is included in the query result but will no longer be valid query result after the update.
3. The object is not included in the query result but will become the query result after the update.
4. The object is not included in the query result and will also not be the query result after the update.

Among these four cases, only the second and third cases influence the query results. In the second case, we need to remove the updated object from the affected query results; while in the third case, we need to add the object to the affected query results. The challenge is how to efficiently categorize each update message into one of the four cases against all existing CPL queries. A brute-force approach that scans all the queries and check if the object is in or will in their query results is obviously time consuming since an object may just affect a small set of existing queries. Therefore, to reduce unnecessary comparisons, we leverage the proposed TPR^Q-tree and propose three query maintenance algorithms with increasing performance achievements: (1) solo-update maintenance; (2) solo-object maintenance; (3) batch-object maintenance. The details of the three maintenance algorithms are presented in the following subsections.

Solo-Update (SU) Maintenance. The solo-update (SU) maintenance algorithm considers the update of an object information as two parts separately: the deletion of the old object information and the insertion of the new object information. Correspondingly, the SU algorithm conducts two searches on the TPR^Q-tree for each object update. The first search looks for a set of CPL queries (denoted as Q_{old}) to which the object belongs to at the object's previous update timestamp t_{old}; the second search looks for a set of CPL queries (denoted as Q_{new}) to which the object belongs to after its update at current timestamp t_{new}. Note that in the case when a new object joins the system (an insertion only), the first search will be skipped and only the second search will be executed. On contrast, in the case when an object exits the system (a deletion only), only the first search will be executed.

To obtain the query set Q_{old}, we start the search from the top of the TPR^Q-tree. For each entry of the visited internal tree node, we compute its MBR at t_{old}. Recall that MBRs and influence regions stored in the TPR^Q-tree are associated with shrinking speed. Therefore, to obtain the MBR at t_{old}, we need to expand it on all the four directions by $MBR_{speed} \cdot (t_u - t_{old})$, where t_u is the last time that the MBR is updated. Then, we check if the old object position falls into the expanded MBR. If so, that means this object may be included in the CPL queries stored under the children leaf nodes of this entry. Therefore, we will further check the children nodes of this entry in the similar way.

When the search reaches the leaf node, we do not simply scan all the query results associated with each query in this node because it could be time consuming. Instead, we take the advantage of the object travel destination and the influence region to prune queries that definitely do not contain the old object position. First, we prune the CPL queries whose query road segments are not on the traveling direction of the object according to its old travel destination. Then we compute the influence regions of the remaining queries at t_{old}. The center of the old influence region is the same as the one stored in the tree, while the inner radius (r_{old_inner}) and the outer radius (r_{old_outer}) are computed based on the inner/outer speed multiplied by the time difference as shown in Eqs. 1 and 2, respectively. If the old object position is within the old influence region of the CPL query, that means this object may be included in the corresponding CPL query. Then, we further check the actual query results of this query and remove the object if found. Moreover, the speed of the influence regions of the affected CPL queries and the MBRs in their ancestor nodes may need to be recalculated if the deleted object contributes to the minimum or maximum shrinking speed.

$$r_{old\_inner} = RoadDist(speed_{min} \cdot (t_q - t_{old})) \qquad (1)$$
$$r_{old\_outer} = RoadDist(speed_{max} \cdot (t_q - t_{old})) \qquad (2)$$

An example of computing Q_{old} is illustrated in Fig. 10 which shows an object's old position (the black circular point), its old destination (denoted as a star) and the influence regions of five CPL queries (a, b, c, d, e) at t_{old}. Queries a, b and c are pruned using the object's old travel destination since they are not in the traveling direction of the object. Then, the influence regions at t_{old} of queries d and e are computed. Since the object is located in both queries' influence regions, the result lists of the two queries will both be checked.

The process for identifying the query set Q_{new} is very similar to that for Q_{old}. The main differences are the computations of the MBRs and the influence regions used during the search. Since t_{new} is after t_u (the latest update time of the MBR), the MBR at t_{new} is computed by shrinking the stored MBR at four

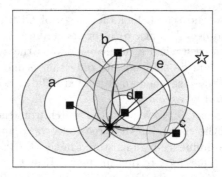

Fig. 10. Influence regions at t_{old} in a leaf node of the TPR^Q-tree

---

Procedure SU Maintenance

Input : $t_{old}, x_{old}, y_{old}, des_{old}, t_{new}, x_{new}, y_{new}, des_{new}, vId$

Output : Q_{new} and Q_{old}

1.   **if** t_{old} is not NULL **then**
2. $Q_{old} \leftarrow isContainPoint(t_{old}, x_{old}, y_{old}, TPR^Q.root)$
3. $DeleteOldResult(Q_{old}, vId)$
4.   **if** t_{new} is not NULL **then**
5. $Q_{new} \leftarrow isContainPoint(t_{new}, x_{new}, y_{new}, TPR^Q.root)$
6. $InsertNewResrult(Q_{new}, vId)$
7.   Report updated query results to the user

Fig. 11. Description of the `Solo-Update Maintenance` algorithm

directions by $MBR_{speed} \cdot (t_{new} - t_u)$. The influence regions of CPL queries at t_{new} is computed based on the following equations.

$$r_{new\_inner} = RoadDist(speed_{min} \cdot (t_q - t_{new})) \qquad (3)$$

$$r_{new\_outer} = RoadDist(speed_{max} \cdot (t_q - t_{new})) \qquad (4)$$

For each CPL query in the obtained Q_{new}, we add the object's new position to its query result. Also, the speeds of the influence regions of the queries in the Q_{new} and the MBRs of their ancestor nodes may need to be updated by considering this object's new speed.

Finally, we record the number of changes for each query result during the object update. If the number exceeds the specified threshold ρ, the server will return the latest query results to the query issuer. An outline of the SU maintenance algorithm is given in Figs. 11 and 12 outlines the algorithm to check the overlap of the point and the node-MBR.

Solo-Object (SO) Maintenance. In general, an object's new and old positions in the same update message are relatively close to one another since they

Procedure isContainPoint()
Input : $t, x, y, destination, root$
Output : Q

1. $node \leftarrow root$
2. $Q \leftarrow empty$
3. $nodeList \leftarrow \{node\}$
4.   **while** $node$ **not** $leafnode$ **do**
5.   **for each** $entry\ ent \in node$ **do**
6. $MBR \leftarrow compute\ ent.MBR\ at\ time\ t$
7.   **if** (x, y) $is\ in\ MBR$ **then**
8. $nodeList \leftarrow nodeList \bigcup \{ent.child\} - \{node\}$
9. $node \leftarrow nodeList[0]$
10.   **while** $nodeList$ **not** $empty$ **do**
11.   **for each** $entry\ ent \in node$ **do**
12.   **if** $ent.CPL\ is\ on\ the\ object's\ destination$ **then**
13. $IR \leftarrow$ compute the influence region of $ent.CPL$ at time t
14.   **if** $(x, y) isin IR$ **then**
15.   **if** (x,y) is included in the $ent.CPL$ **then**
16. $Q \leftarrow Q \bigcup ent$
17.   **return** Q

Fig. 12. Description of the `isContainPoints()` algorithm

are two consecutive positions on the object's path and bounded by the maximum moving speed multiplied by the maximum update interval. Therefore, the new and old positions in an object's single update message are very likely to be covered by the influence regions of the same or nearby CPL queries. In other words, these two positions may affect the CPL queries stored in the same or sibling nodes in the TPR^Q-tree. Based on this observation, we propose the solo-object (SO) maintenance algorithm that considers the object update message as a whole and computes the two sets of CPL queries affected by the update (i.e., Q_{new} and Q_{old}) simultaneously in one round of the search in the TPR^Q-tree. The SO algorithm is expected to be more efficient than the previous discussed SU algorithm because the SU algorithm carries out two rounds of the search separately for the new and old positions, which may visit the same tree nodes repeatedly.

Figure 13 presents an outline of the SO maintenance strategy. In particular, we start the search from the root of the TPR^Q-tree. For each entry of the visited internal node, we compute its MBRs at t_{old} and t_{new}, respectively, in the similar way as discussed in the SU algorithm. If the object's old or new position is covered by the MBRs, the child node of this entry will be added for checking as well. Until the leaf level is reached, the influence regions of the CPL queries stored in the visited entries will be computed at t_{old} and t_{new} respectively. Then, the old and new positions will be compared against the respective influence regions. If the old position is included in the influence region of a CPL query,

Procedure SO Maintenance
Input : $t_{old}, x_{old}, y_{old}, t_{new}, x_{new}, y_{new}, vId$

1. $Q_{old} \leftarrow empty$
2. $Q_{new} \leftarrow empty$
3. $node \leftarrow root$
4. $nodeList \leftarrow \{node\}$
5.   **while** $node$ **not** $leaf node$ **do**
6.   **for each** entry ent of the node **do**
7. $MBR_{old} \leftarrow compute e.MBR at time t_{old}$
8. $MBR_{new} \leftarrow compute e.MBR at time t_{new}$
9.   **if** $(old_x, old_y) is in MBR_{old}$ **then**
10. $nodeList \leftarrow nodeList \bigcup \{ent.child\}$
11.   **else if** (new_x, new_y) is in MBR_{new} **then**
12. $nodeList \leftarrow nodeList \bigcup \{ent.child\}$
13.   remove node from $nodeList$
14. $node \leftarrow nodeList[0]$ \\get the first node in the nodeList
15. $NodeUpdateList \leftarrow empty$
16.   **while** $nodeList$ is not empty **do** \\now check the leaf nodes
17.   **for** each entry ent of the node **do**
18.   **if** $ent.CPL$ is on the object's old destination **then**
19. $IR_{old} \leftarrow$ compute the influence region of $ent.CPL$ at time t_{old}
20.   **if** (old_x, old_y) is in IR_{old} **then**
21.   **if** (old_x, old_y) is included in the $ent.CPL$ **then**
22.   remove (old_x, old_y) from $ent.CPL$
23. $NodeUpdateList \leftarrow NodeUpdateList \cup ent$
24.   **else if** $ent.CPL$ is on the object's new destination **then**
25. $IR_{new} \leftarrow$ compute the influence region of $ent.CPL$ at time t_{new}
26.   **if** (new_x, new_y) is in IR_{new} **then**
27.   **if** (new_x, new_y) are the new answer to $ent.CPL$ **then**
28.   add (old_x, old_y) to $ent.CPL$
29. $NodeUpdateList \leftarrow NodeUpdateList \cup ent$
30.   Recalculate the IR of entries in $NodeUpdateList$
31.   Update the MBRs of the ancestor nodes of entries in $NodeUpdateList$
32.   Report updated query results to the user

Fig. 13. Description of the `Solo-Object Maintenance` algorithm

the old position will be removed from the query result. If the new position contributes to a CPL query, the new position will be inserted to the query result. Next, the shrinking speeds of influence regions of all the updated CPL queries will be recalculated. The MBRs of the ancestors of the updated entries will be recomputed as well. At the end, if the query results have been changed significantly (exceeding certain threshold), a query update report will be sent back to the query issuers.

Batch-Object Maintenance. With the increase of the number of moving objects, the number of object updates at each timestamp will also grow larger.

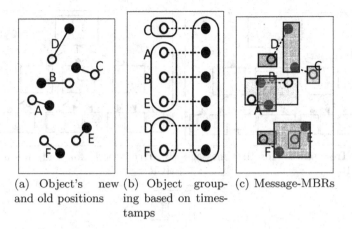

(a) Object's new and old positions

(b) Object grouping based on timestamps

(c) Message-MBRs

Fig. 14. Group formation for a set of update messages

Among a large amount of updates that received at the same timestamp, it is likely that some are from nearby objects and hence they may influence the same or nearby CPL queries. According to this observation, we take one step further from the previous SO algorithm by considering all updates received at one single timestamp as a whole, and propose a batch-object (BO) maintenance algorithm.

Upon receiving the update messages at a timestamp, the BO algorithm first conducts two rounds of grouping: (i) grouping objects based on their update timestamps; (ii) grouping objects based on their location proximity. In the first round of grouping, the objects' new positions can be easily grouped together as they are all at the same current timestamp. The challenging design issue is the grouping of the objects' old information. This is because objects which issue updates at the same time now may have issued their last updates at totally different timestamps. In other words, the different timestamps associated with the old positions make these old positions incomparable. We cannot directly group the old positions based on only location proximity but overlooking their update timestamps. To overcome this problem, we group old positions based on their update timestamps by putting the old positions with the same update timestamp into the same group. So far, we obtain one group for the objects' current positions and multiple groups for the objects' old positions. The benefit of the first round of grouping is that it avoids repeated computation of the MBRs and influence regions in the TPR^Q-tree for objects falling into the same timestamp. Next, we divide the obtained groups into sub-groups based on the location proximity. Specifically, we employ the similar technique in the R*-tree by constructing MBRs for the nearby objects. For the clarify of the subsequent discussion, we call the MBRs constructed from update messages, the *message-MBRs*.

Figure 14 illustrates the group formation for a set of object-update messages. In Fig. 14(a), the circles denote the old positions and the black points denote the new positions of six objects: A, B, C, D, E, and F. All these update messages were received at time t_{new}. The previous updates of objects A, B, C, D, E and

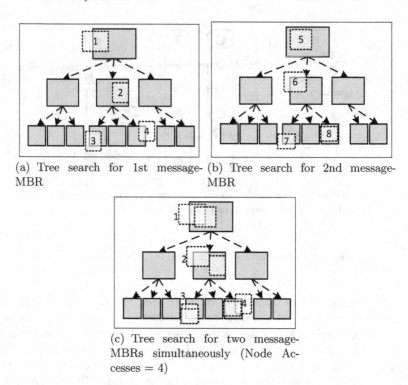

(a) Tree search for 1st message-MBR

(b) Tree search for 2nd message-MBR

(c) Tree search for two message-MBRs simultaneously (Node Accesses = 4)

Fig. 15. Different strategies for searching the message-MBRs

F were made at time t_1, t_1, t_0, t_3, t_1, and t_3, respectively. Figure 14(b) shows the update messages grouped according to their timestamps. As shown in the figure, new positions form a single group as they all have the same timestamp. Old information, however, forms three different groups with object C in the first group, objects A, B and E in the second group, and objects D and F in the third group. Then, Fig. 14(c) shows the message-MBRs of further partitioning of the three groups based on their location proximity.

After the grouping, the next step of the SO algorithm is to search the TPR^Q-tree to find the CPL queries that overlap with the message-MBRs, i.e., to find the CPL queries that may be affected by this set of object updates. Here, if we search each message-MBR in the TPR^Q-tree, there will still be repeated accesses to the same tree node. For example, suppose that the received update messages form two message-MBRs. Figure 15(a) and (b) show the search of the first and the second message-MBRs respectively, where the dashed rectangles denote the message-MBRs and the number is the count of the page accesses). As shown, the two searches accessed the same tree nodes consecutively and result in 8 total node accesses. We can observe that if the two searches are carried out simultaneously as shown in Fig. 15(c), the repeated node accesses can be avoided and the cost will be cut in half. Therefore, in our BO algorithm, we consider all

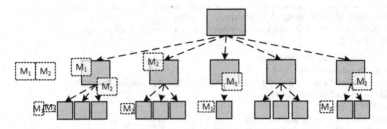

Fig. 16. Message-MBRs overlapping with MBRs in the TPR^Q-tree

message-MBRs against the MBR in the same tree node to ensure that each tree node is accessed at most once for a set of updates received at the same timestamp.

Moreover, we also notice that not all message-MBRs overlap with the MBR of the examined tree node. Figure 16 illustrates this kind of situation, whereby the two message-MBRs M_1 and M_2 overlap with different nodes in the TPR^Q-tree. If a message-MBR does not overlap with the MBR of a node in the TPR^Q-tree, there is no need to further consider this message-MBR under the branches of this tree node. Our BO algorithm leverages this pruning criteria which greatly reduces the amount of comparison as well as the computation of the MBRs and influence regions needed for the comparison.

An overview of the BO algorithm is shown in Fig. 17. First, the message-MBRs are obtained (line 2–3). Then, the search starts from the root of the TPR^Q-tree (line 4). For each visited internal node, we check the flags of all the message-MBRs to see if this node's parent overlaps with the message-MBRs. If so, we further compare this message-MBR with the MBRs of each entry in the examined node. Flags are updated for the children node of each entry after the comparison (line 13). As for the leaf (line 16–27) node, we also check the flags first. For the candidate CPL queries obtained from the search, we finally evaluate the query against the actual object position in the update message to adjust the query results similarly to that in the previous two maintenance algorithms.

6 Query Cost Analysis

The CPL query algorithms consist of two phases: the initial phase and the maintenance phase. At the initial phase, a snapshot predictive line query is executed and the cost of this snapshot query has been analyzed in [7]. Moreover, all the proposed three algorithms share the same initial phase while differ in the maintenance phase. Therefore, we focus on the analysis of the maintenance cost in this section.

In our cost analysis, we assume that both moving objects and querying road segments are uniformly distributed in the space being considered. Without loss of generality, we also assume that all moving objects are alive during the life time of the queries and all the queries considered are issued at the same timestamp

Procedure BO Maintenance Algorithm
Inputs : *updates*: a set of update messages received at the same time stamp
Outputs : Q_{Result}

1. $G \leftarrow$ groups of object updates at the same timestamp
2.   **for each**   group G **do**
3.   message-MBR $\leftarrow$ group objects in G according to location proximity
4. $node \leftarrow$ root of the TPR^Q-tree
5. $nodelist \leftarrow \{node\}$
6.   **while** $node$ is not the leafnode **do**
7.   **for each**   message-MBR **do**
8.   **if** flag($node$, message-MBR) is true **then**
                        \\this node's parent overlaps with the message MBR
9.   **for each**   entry in $node$ **do**
10.   compute the MBR at the message-MBR's timestamp
11.   **if** MBR overlaps with the message-MBR   **then**
12.   add this entry's child node to the NodeList
13.   set the $flag(entry.child$, message-MBR) to *true*
14.   remove node from *nodeList*
15. $node \leftarrow nodeList[0]$ \\get the first node in the nodeList
16.   **while** *nodeList* is not empty **do**
17.   **for each**   entry in the *node* **do**
18.   **for each**   message-MBR **do**
19.   **if** $flag(node$, message-MBR) is *true* **then**
20.   compute the IR at the message-MBR's timestamp
21.   **for each** *position* contributes in message-MBR **do**
22.   **if** the *position* is in IR **and** *position* is included in the *ent.CPL* **then**
23.   **if** message-MBR's timestamp is the new timestamp **then**
24.   add *position* to *ent.CPL*
25.   **else**
26.   remove *position* from *ent.CPL*
27. $NodeUpdateList \leftarrow NodeUpdateList \cup ent$
28.   Recalculate the IR of entries in $NodeUpdateList$
29.   Update the MBRs of the ancestor nodes of entries in $NodeUpdateList$
30.   Report updated query results to the user

Fig. 17. Description of the `Batch-Object Maintenance` algorithm

with the same length of life time. We estimate the average maintenance cost in terms of the number node accesses (or disk page accesses assuming one node per disk page). Specifically, the average maintenance cost per query per timestamp is computed as the total number of disk page accesses ($Cost_{total}$) divided by the product of the total number of CPL queries (N_q) and the total timestamps (T) during the query life time (Eq. 5).

$$Cost = \frac{Cost_{total}}{N_q \cdot T} \tag{5}$$

6.1   Cost of Solo-Update (SU) Maintenance

To obtain the average maintenance cost according to Eq. 5, we only need to estimate the unknown value, i.e., $Cost_{total}$. The total number of page accesses ($Cost_{total}$) during the query life time using the SU algorithm is the multiplication of two factors: the number of times that the TPR^Q-tree is accessed and the number of page accesses per tree access.

The number of times that the TPR^Q-tree is accessed is twice of the total number of update messages in the system. This is because the SU maintenance approach treats one update message as a deletion followed by an insertion. Let m_i denote the total number of update messages from an object i during the query life time T. The total number of update messages in the system is computed as $\Sigma_{i=1}^{N}(m_i)$, where N is the total number of objects. Then, the total number of tree accesses (denoted as $Total_{ta}$) by the SU algorithm is $2\Sigma_{i=1}^{N}(m_i)$.

The second step is to estimate the cost of searching the TPR^Q-tree for a single operation (either a deletion or an insertion). Given an object's old or new position, the average number of CPL queries whose influence regions may contain this position is determined by the area of the influence region at the update timestamp and the density of the queries in the space being considered. The area covered by the outer circle of an influence region at the query update timestamp t_u is estimated as $\Pi \cdot [(T-t_u) \cdot SpeedMax]^2$, where $(T-t_u) \cdot SpeedMax$ is the outer radius r_{out} of the influence region at t_u. Since t_u evolves from time 0 to T, the average area of the influence region (denoted as $Area_{IR}$) during T is the integration $\frac{\int_0^T \Pi \cdot r_{out} dt}{T}$ which is equal to the following:

$$Area_{IR} = \frac{\Pi \cdot (SpeedMax)^2 \cdot T^2}{3}$$

Take the object position as the center and draw a circle of the size of $Area_{IR}$ as illustrated in Fig. 18, where the dark point in the center represents the object O's position and the circles drawn in solid lines represent the outer circle of the CPL queries' influence regions. If a CPL query's querying road segment intersects with the object's circle, this query's influence region may contain the object. In other words, the CPL query whose querying road segments in the shaded area should be considered.

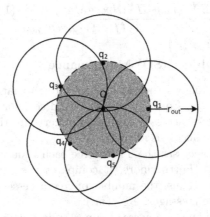

Fig. 18. Example of the CPL queries that may contain the object

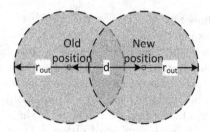

Fig. 19. Maximum query overlapping area corresponding to one update message

Next, we estimate the number of queries that may fall into the object's influence circle. Assuming the road segments being queried are uniformly distributed, the number of CPL queries per unit area is $Density_{query} = N_q/Area_{total}$. Thus, the number of queries in the object's influence circle is the multiplication of the area of the influence circle and the density, which is $n_q = Area_{IR} \cdot Density_{query}$. Since these n_q queries are close to one another, they are likely to be stored close to one another in the TPR^Q-tree as well. Therefore, the average number of leaf node containing the n_q queries is estimated as n_q/f, where f is the fanout (i.e., average number of entries per node) of the TPR^Q-tree. The number of the parent nodes of the leaf nodes containing these n_q queries is estimated as n_q/f^2. In general, the number of nodes accessed at the level l of the tree is e (n_q/f^l), where the level of the leaf node is 1. After summing up the node accesses at each level, we obtain the total number of node accesses during one round of the tree search: $C_{tree} = \sum_{l=1}^{h}(n_q/(f^l))$, where h is the height of the tree.

Finally, we can compute the average query cost of the SU maintenance as follows:

$$Cost_{su} = \frac{Total_{ta} \cdot C_{tree}}{N_q} \tag{6}$$

$$= \frac{2\Sigma_{i=1}^{N}(m_i) \cdot \Pi \cdot (SpeedMax) \cdot T)^2 \cdot (1 - (1/f^h))}{3 \cdot (f - 1) \cdot Area_{Total}} \tag{7}$$

6.2   Cost of Solo-Object (SO) Maintenance

The SO cost analysis follows the same procedure as that of the SU. The SO algorithm also utilizes the Eq. 5. To find the total cost for the entire query life time, the number of tree accesses during the query life time and the average page accesses for a tree access is estimated.

The SO algorithm accesses the TPR^Q-tree each time it received an update message (refer Fig. 13). Furthermore, it compares the entire update message against the TPR^Q-tree. Thus, the number of tree accesses in SO is simply the total number of update messages: $\Sigma_{i=1}^{N}(m_i)$.

The search cost for one tree access (or per update message, in other words) for SO algorithm is also estimated as it was for SU algorithm. Thus, we first find the area that the object may influence query results. This area is illustrated in the Fig. 19. The small circles at the center of the bigger circles are the object's new and old positions. The bigger circles represent the maximum query overlap area of each position. The distance between two positions is d. The total query-affected area is the area covered by the boundaries of the two circles: $(2\Pi \cdot r_{out}^2) - (r_{out}^2 \cdot \arccos(\frac{d}{2r_{out}}))$ and the average query-affected area over the T time period is $Area_{IR,SO} = \frac{\int_0^T (2\Pi \cdot r_{out}^2) - (r_{out}^2 \cdot \arccos(\frac{d}{2 \cdot r_{out}}))\mathrm{d}t}{T}$, which is equivalent to:

$$[2 \cdot \Pi \cdot s^2 \cdot T^2] - [\frac{s^2 \cdot T^2 \cdot \arccos(\frac{d}{2 \cdot s \cdot T})}{2}]$$
$$+[\frac{d}{6s^2T} \cdot (\frac{(u_T^{1.5} - u_0^{1.5})}{3} + \frac{d_2 \cdot (u_T^{0.5} - u_0^{0.5})}{4})]$$

where s is the $SpeedMax$, and $u_i = (SpeedMax \cdot t)^2 - (\frac{d^2}{4}); i \in \{0, T\}$.

The number of queries in the influence region $n_{q,SO}$ is $Area_{IR,SO} \cdot Density_{query}$. Following the same procedure as in SU cost analysis, the total number of node accesses during one round of tree search $C_{tree,SO}$ is obtained as: $C_{tree,SO} = \sum_{l=1}^{h}(n_q/(f^l))$ (h is the height of the tree). Then the average query cost of SU maintenance becomes:

$$Cost_{so} = \frac{\sum_{i=1}^{N}(m_i) \cdot Area_{IR,SO} \cdot (1 - (1/f^h))}{(f-1) \cdot Area_{Total}} \tag{8}$$

Theorem 1. *The maintenance cost of the SO algorithm is always no greater than the cost of the SU algorithm.*

Proof. The worst case of the SO algorithm is obtained when each point accesses entirely different tree nodes. This means that no overlap between the circles showed in Fig. 19 exists. When there is no overlap between circles, d becomes zero and the area covered by two circles (i.e., $Area_{IR,SO}$) becomes $(2\Pi \cdot r_{out}^2)$. When the value of $Area_{IR,SO}$ is plugged on Eq. 8, it is simplified to Eq. 7, which is the cost for SU algorithm. □

6.3 Cost of Batch-Object (BO) Maintenance

The BO algorithm also needs to find the number of tree accesses and the number of page accesses per each tree access to estimate the $Cost_{total}$ in Eq. 5.

The number of tree accesses in BO algorithm depends on the number of distinct timestamps at which update messages are initiated. Because, the BO algorithm groups update messages received at the same timestamp and access the tree only once per all messages in the same timestamp. Thus, assuming the number of distinct update message timestamps are N_{ts}, the TPR^Q-tree accesses is also the N_{ts}.

The average page accesses per each tree access depends on the number of subgroups and their MBR extent. A subgroup whose MBR dimensions are

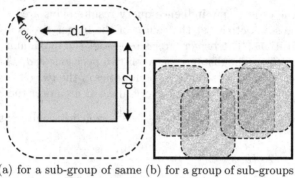

(a) for a sub-group of same (b) for a group of sub-groups
time stamped positions

Fig. 20. Maximum Query-overlap area

d_1 x d_2 and its maximum query overlap area are shown in the Fig. 20(a). The
MBR of the subgroup is represented by the filled rectangle and the maximum
distance to a query road segment from the MBR boundary is the r_{out}. The area
covered by the dash-lined shape is the influence region of the MBR. Its area is
calculated as follows:

$$a_i = (d_1 \cdot d_2) + 2(d_{i1} \cdot r_{out}) + 2(d_{i2} \cdot r_{out}) + \Pi r_{out}^2. \qquad (9)$$

Since all MBRs are compared simultaneously against each tree node, repet-
itive node accesses are not counted. The page accesses per one tree-search is
the total distinct node accesses on the TPR^Q-tree. This means that the common areas in different query influence regions should be counted only once.
Figure 20(b) shows an example of overlapped query-influence areas. This area is
given in Eq. 10.

$$\sum_{i=0}^{n-1} a_i - \sum_{i=0}^{n-2} \sum_{j=i+1}^{n-1} Overlap_{i,j} \qquad (10)$$

The answer to this calculation is approximated to the area of the MBR
covered by all round-cornered rectangles $Area_{IR,BO}$ (see Fig. 20(b)). Then the
average number of queries that can overlap with the area is:

$$n_{q,BO} = Area_{IR,BO} \cdot Density_{query} \qquad (11)$$

Following the same cost estimation steps as in the SU and SO cost analysis,
the average BO maintenance cost becomes:

$$Cost_{BO} = \frac{N_{ts} \cdot Area_{updateMBR} \cdot (1 - (1/f^h))}{(f-1) \cdot Area_{Total}} \qquad (12)$$

Theorem 2. *The maintenance cost of the BO algorithm is always less than the
cost of the SO algorithm.*

Proof. The worst case of MO algorithm is obtained, when each MBR accesses distinct tree nodes. To have distinct node accesses, no overlap should exist among MBRs. This can be explained with Eq. 10. According to the equation, when $Overlap_{i,j}$ $\forall i,j$ is zero, the maximum affective area is obtained and it is simply the summation of all MBRs areas.

Then, let us consider number of elements in a group is $elements_g$. Hence, N_{ts} in Eq. 12 can be re-written as $Total_{messages}/elements_g$. The number of sub-groups also can be obtained in terms of $elements_g$. For that, assume the average number of elements in a subgroup to be $elements_{sg}$. Then, the average number of subgroups $n = 1 + (elements_g/elements_{sg})$. The maximum aggregated area of MBRs is obtained when n is large. The largest n is obtained when $elements_g$ is largest and $elements_{sg}$ is the smallest. The smallest possible $elements_{sg}$ is one. When, $elements_{sg}$ is one, each a_i in Eq. 10 becomes Πr_{out}^2, according to Eq. 9. With the deduced parameter values Eq. 12 can be simplified as follows:

$$Cost_{BO} = \frac{\frac{Total_{messages}}{elements_g} \cdot \sum_{i=0}^{elements_g+1} \Pi r_{out}^2 \cdot (1 - (1/f^h))}{(f-1) \cdot Area_{Total}}$$

In this equation, when $elements_g = 1$, cost for SO algorithm is obtained. To sum up, the BO obtain the SO maintenance cost, when: no overlaps between subgroups are exist, number of objects in one group is one, and number of elements in sub groups is one (i.e., two subgroups in the group).   □

7   Performance Study

The proposed algorithms were evaluated on moving object data sets generated by the Brinkhoff' generator [1]. The moving object datasets were generated using four real road maps selected from different states in United States. The road maps contain a similar number of road segments, but different topologies. The number of moving objects in each dataset ranges from 10k to 100k.

For each dataset, sets of queries were randomly generated by randomly select-ing query issuer and its query issuing position. Then the querying road segment was selected from its path which will be reached by the end of the predictive query length. The predictive query lengths were ranged from 10 min to 60 min. The chosen parameters and their values are presented in Table 1. The bold values represents the default value for each parameter.

We compare our proposed approaches with a naive approach that executes snapshot predictive line queries [7] for every update message. Since the initial phase of the four approaches are the same, in the following, we only report the comparison of their maintenance cost. The performance is measured in terms of the prediction error rate and the I/O cost. The error rate was computed by comparing the number of objects in the predictive query results with the actual number of objects on the query road segment at the query time. The I/O cost is the number of disk page accesses. The reported I/O cost is the average page

Table 1. Simulation parameters and their values

Parameters	Values
Buffer	YES, **NO**
Number of queries	0.5 %, 2 %, 5 %, **20%**, 40 %, 60 %, 80 %, 100 % 20 K
Number of moving objects	10K, 20K, . . . , **50K**, 60K, . . . , 100K
Predictive time length	10, 20, **30**, 40, 50, 60 (mins)
Road maps	Alpine (CA), Charles (MD), Salem (NJ), **Worth (MO)**

accesses per query per timstamp. It first calculates the average page accesses (averaged per query and per timestamp) during each 5 mins time interval though out the query life time ($AvgPg(5min)$). The average page accesses for the entire query life time is, then, calculated by taking the average of all $AvgPg(5min)$'s in the query life time.

7.1 Maintenance Cost of CPL Queries

In the following, we evaluate various factors that may affect the query performance including the time, the number of queries, the number of moving objects, the predictive length, the road topology, and the buffer size.

Query Performance over the Query Life Time. First, we evaluate the performance of the query result maintenance as time passes. We compute the average maintenance cost and prediction error rate per timestamp within each 5-minute interval for 30 min. Figure 21(a) and (b) report the performance of the naive approach and the three proposed approaches: Solo-Update (SU), Solo-Object (SO), and Batch-Object (BO).

From Fig. 21(a), we can observe that our proposed three algorithms consistently yield much lower prediction error than the naive approach. This is because the naive approach defines the query ring based on the Euclidean distance to the query road segment [7], whereas the influence regions employed by SU, SO and BO consider the road distance which is more accurate to estimate vehicles that may enter the query road segment. In addition, we can also see that the prediction accuracy of our three algorithms is similar which is not affected by the various maintenance algorithms adopted.

Figure 21(b) shows the average maintenance cost. As expected, our proposed three algorithms all perform better than the naive approach, and the BO approach performs best. This is because the naive approach needs to execute each query every timestamp which may involve duplicate efforts when there is no change to the results. The TPR^Q-tree, which is utilized by the proposed algorithms, takes object update messages and checks all affected queries simultaneously, which helps reduce the unnecessary efforts on the query processing significantly. The reason that the BO approach achieves the least maintenance cost is that the BO approach is the most aggressive one among the three proposed

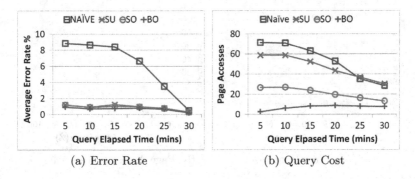

(a) Error Rate   (b) Query Cost

Fig. 21. Query performance over the query life time

approaches and considers most possibilities of simultaneous executions for cost saving. In addition, the experimental results also demonstrate the evolvement of the maintenance cost with time. As shown in the figure, the closer to the end of the query life time (i.e., the time when the query issuer will enter the querying road segment), the less maintenance cost is needed in general. The possible reason of such behavior is that the influence regions are shrinking as time passes and hence the number of objects to be checked become fewer. Note that the BO approach shows a slight increase of the maintenance cost at the beginning. The reason is the following, the BO approach considers all the updates issued at one timestamp and the number of updates are fewer when the system just starts because the objects take some time to speed up.

Effect of the Number of Queries. In this round of experiments, we evaluate the effect of the number of queries on the query performance by varying the total number of queries from 0.5 % of the total number of moving objects from 100 %. As shown in Fig. 22, the naive approach exhibits a relatively stable performance regardless of the number of queries. This means that the average cost per one query is independent from the total number of queries being executing. Each query is applied on the same process and on the same tree (R^D-tree). Hence, the cost depends only on the size of the R^D-tree, but not the number of queries. Our proposed SU, SO, and BO approaches, however, access the TPR^Q-tree and the number of queries stored in the tree changes the tree structure. In fact, the number of queries decides the tree fanout (f) and the height of the tree (h). These two factors directly impact on the query maintenance cost. The query cost, in all three proposed algorithms, is proportional to the expression $\frac{1-\frac{1}{f^h}}{f-1}$. The impact of h and f is contravened on both the this expression and the average query cost.

For smaller h values, impact of both f and h is significant. For example, for 0.5 % (250 in count) of queries, all queries can be accommodated in the root; which means h is one and f is greater (refer Table 2). When the number of queries is increased up to 2 %, the number of tree levels increases and, at the same

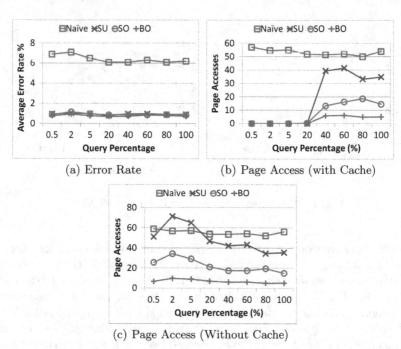

(a) Error Rate   (b) Page Access (with Cache)

(c) Page Access (Without Cache)

Fig. 22. Effect of number of queries

Table 2. TPRQ-tree structure's information

Query percentage	0.5 %	2 %	5 %	20 %	40 %	60 %	80 %	100 %
Number of queries	250	1000	2500	10000	20000	30000	40000	50000
Number of tree levels	1	2	2	2	2	2	2	3
fanout	231	182	159	180	180	174	180	179

time, fanout decreases. Both these changes result to increase the value of the expression. When the h gets bigger, the expression becomes nearly independent of h as $\frac{1}{f^L}$ becomes insignificant. The expression is, then, left only to f. Hence, as the number of levels is increased (i.e., higher number of queries), a smooth query cost decrement is demonstrated.

Effect of Buffer Utilization. We repeat the set of experiments conducted in the previous section to see the effect of the buffer utilization. Specifically, we employ a buffer with 50k capacity and LRU (Least Recently Used) replacement policy. Figure 22(b) reports the query cost for deferent query percentages with the buffer[3]. As the figure shows, the query maintenance cost up to 20 % is essentially a zero and the rest of the query sets shows an increased query cost.

---

[3] Since the accuracy is not affected by the buffer, it is omitted in the discussion.

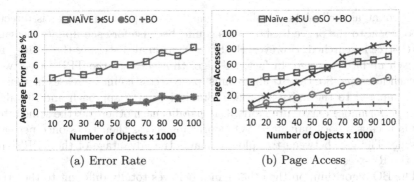

(a) Error Rate   (b) Page Access

Fig. 23. Effect of number of objects

The increased costs are comparable to that of in Fig. 22(c). Comparing Fig. 22(c) with Fig. 22(b), it is clear that the query performance up to 20 % has improved due to the buffer usage. This is because, up to 20 %, the number of tree nodes in the entire tree structure is less than 50. This means that the entire tree can be accommodated by the buffer. Thus, at most one disk access is made per one tree node. Once the node is stored in the buffer, no buffer replacement is required. When the number of nodes in the tree exceeds the buffer size, buffer cannot accommodate all necessary tree nodes simultaneously. Thus, buffer-miss rate increases and hence page access count increases.

Effect of Number of Moving Objects. In this round of experiments, we evaluate the performance when the number of moving objects increases from 10K to 100K. Figure 23(a) shows the average error rate of proposed algorithms together with the naive approach. As the figure shows, similar to the other cases reported in early sections, all three algorithms show competitive accuracy. The error rates, in all approaches, increase slightly with the number of objects. This is because more the moving objects, more the uncertainty of the prediction. However, our approach always achieves a lower error rate for the same reason discussed in the previous section.

Figure 23(b) shows the query cost of all four algorithms. According to the graph, one common observation on all algorithms is they all consume more page accesses when the object count is increased. In the naive approach, this happens because the R^D-tree expands with the higher number of objects and hence the number of node accesses is increased. In the proposed algorithms, the tree structure remains unchanged, but the number of update messages compared against the tree is increased. Another vital observation is the naive approach gives the worst query cost for lesser number of objects, and it defeats the performance of SU when the number of objects are increased (approximately at 60k). The reason can be explained as follows. The page access count in naive approach depends on two factors: the size of the R^D-tree and the number of update messages received. The expansion of R^D-tree is slower for higher object counts than the smaller object counts. This same expansion speed will be applied on the page access count as well. Additionally, the number of update messages is directly proportional to the pageedrxsh0 access count, because for each update message, the R^D-tree is searched. However, the SU algorithm also accesses the TPR^Q-tree per each update message. In fact, SU algorithm accesses the TPR^Q-tree twice per each message. So, the SU algorithms' page access count increases in a faster rate compared to the naive approach. Similarly, naive approach and SO algorithm performance curves are more likely parallel each other (i.e. the same rate). This is because, both naive and SO algorithms access their trees once per each message. The gap between two plots explains the advantage of the TPR^Q-tree over the R^D-tree.

The BO algorithm, on the other hand, behaves totally different to the other approaches and shows extremely better performance. As the figure shows, the BO algorithm has not been affected by the number of object as it was in the other three algorithms, especially when the number of objects is higher. As a matter of fact, the BO algorithm's performance depends on only the number of different time stamps and it is countably finite, within the 30 min time period. Thus, the BO shows a bounded query cost independent of the number of objects.

Effect of Predictive Time Length. In this set of experiments, the predictive time length is varied from 10 min to 60 min. As shown in Fig. 24(a) the error rate stays in a similar range regardless of the predictive time length for both approaches. The behavior can be explained as follows. For the naive approach, it executes the query every timestamp and hence any change of object travel plan will be captured. Similarly, in proposed approaches, the effect of the object update on the query results at every timestamp is considered.

On the other hand, the predictive time length does affect the query cost as shown in Fig. 24(b). The query cost of naive approach increases when the predictive time length is longer. This is because in the naive approach, a bigger ring query is generated for a longer predictive time length. In the proposed approaches also the query cost increases with the length of the query window; but, with a slower rate. This is again due to the advantage of the TPR^Q-tree utilization.

As it was explained in the Sect. 6, the proposed algorithms total query cost depends on either the number of update messages (for SU and SO) or the number of different time stamps within the query life time (for BO). The average query cost for 5 mins time interval depends on the message counts within the 5 mins. Thus, no matter how long the predictive query window is, average query cost depends on the average number of messages within the query window. Given a fixed number of objects (and assuming the same mobile patterns for any query window size) the average number of messages independent on the query window. The other factor that can affect the query cost of proposed algorithms is the query influence area: higher the query window higher the query effective area. Thus, all three proposed approaches experienced slightly higher query cost with the wider query window.

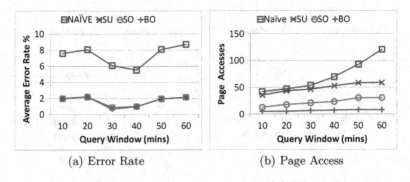

(a) Error Rate   (b) Page Access

Fig. 24. Effect of predictive time length

Effect of Road Topology. This section evaluates the effect of the road topology by testing different maps: Alpine (CA), Charles (MD), Salem (NJ), and Worth (MO). The number of edges in each map was 1576, 1766, 1789, and 1573 respectively, and the average road segment length is 232 m, 370 m, 515 m, and 551 m, respectively. By observing the average error rate of individual topology in Fig. 25(a), it is tend to conclude that the larger the number of edges, the lower the error rate. Regarding the page accesses as shown in Fig. 25(b), our approach is relatively independent of the number of edges. However, all three algorithms show better performance when the average road segment length is bigger. This is because, when the road segments are lengthier, the update messages time interval is spacer. Thus, algorithms handle less update messages.

7.2 Cost Model Evaluation

This section validates the cost model discussed in Sect. 6 for maintenance cost of the proposed three algorithms. The evaluation was performed based on the Eqs. 7, 8, and 12. Figure 26 compares the estimated cost computed from the cost model with the experimental results obtained from the proposed three

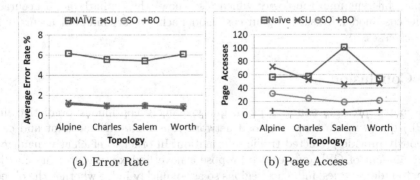

(a) Error Rate   (b) Page Access

Fig. 25. Effect of road topology

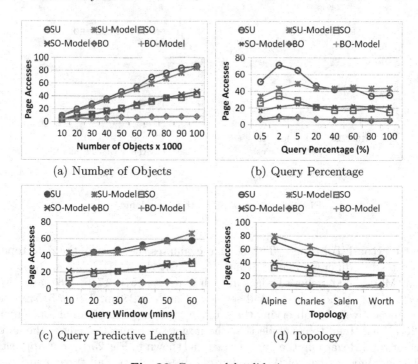

Fig. 26. Cost model validation

maintenance algorithms. Figure 26(a) shows the effect of number of objects. In this case, the cost model's error rate is below 10 %. Figure 26(b) shows the effect of the number queries, whereby the estimation is getting close to the actual cost with the increase of the number of queries. The reason is straightforward. The cost model is developed based on uniform distribution of queries and when there more queries, their distribution is closer to uniform distribution. Next, Fig. 26(c) shows the comparison of the estimated cost and the actual cost in the case when the predictive query length is varied. Again, we can see that the cost model yields an error around 10 %. Finally, Fig. 26(d) reports the comparison results when testing different map topologies which also shows the similarly good accuracy of the cost model. To sum up, our cost model achieves around 90 % accuracy in various cases.

8   Conclusion

This paper presents a new type of query, namely continuous predictive line (CPL) query, which takes the road network constraints into account and continuously provide predicted traffic information. In order to efficiently manage a large amount of CPL queries, we propose a novel index structure, the TPR^Q-tree, to index queries' influence regions so as to quickly judge whether the object updates may affect the continuous query results. Leveraging the TPR^Q-tree, we develop three query algorithms with increasing efficiency on the query performance. We evaluate our approach both theoretically and experimentally, and the results demonstrate the efficiency and effectiveness of our approach.

Acknowledgement. This work is partly funded by the U.S. National Science Foundation under Grant No. CNS-1250327.

References

1. Brinkhoff, T.: A framework for generating network-based moving objects. GeoInformatica **6**, 153–180 (2004)
2. Cai, Y., Hua, K.A., Cao, G.: Processing range-monitoring queries on heterogeneous mobile objects. In: IEEE International Conference on Mobile Data Management (2004)
3. Chen, S., Ooi, B.C., Tan, K.-L., Nascimento, M.A.: ST2B-tree: a self-tunable spatio-temporal B+-tree index for moving objects. In: Proceedings of the 2008 ACM SIGMOD International Conference on Management of Data, SIGMOD 2008, pp. 29–42. ACM, New York (2008)
4. Predic, B., Papadopoulos, A.N., Stojanovic, D., Djordjevic-Kajan, S., Nanopoulos, A.: Continuous range query processing for network constrained mobile objects. In: 8th International Conference on Enterprise Information Systems (2006)
5. Feng, J., Lu, J., Zhu, Y., Mukai, N., Watanabe, T.: Indexing of moving objects on road network using composite structure. In: Apolloni, B., Howlett, R.J., Jain, L. (eds.) KES 2007/WIRN 2007, Part II. LNCS (LNAI), vol. 4693, pp. 1097–1104. Springer, Heidelberg (2007)
6. Gedik, B., Liu, L.: MobiEyes: a distributed location monitoring service using moving location queries. IEEE Trans. Mob. Comput. **5**, 1384–1402 (2006)
7. Heendaliya, L., Lin, D., Hurson, A.R.: Predictive line queries for traffic forecasting. In: Database and Expert Systems Applications (2012)
8. Heendaliya, L., Lin, D., Hurson, A.: Continuous predictive line queries under road-network constraints. In: Decker, H., Lhotská, L., Link, S., Basl, J., Tjoa, A.M. (eds.) DEXA 2013, Part II. LNCS, vol. 8056, pp. 228–242. Springer, Heidelberg (2013)
9. Hu, H., Xu, J., Lee, D.L.: A generic framework for monitoring continuous spatial queries over moving objects. In: Proceedings of the ACM SIGMOD International Conference on Management of Data (2005)
10. Jensen, C.S., Lin, D., Ooi, B.C.: Query and update efficient B^+-tree based indexing of moving objects. In: Proceedings of the 30th International Conference on Very Large Data Bases, VLDB 2004, vol. 30, pp. 768–779. VLDB Endowment (2004)
11. Jensen, C.S., Lin, D., Beng, C.O., Zhang, R.: Effective density queries on continuously moving objects. In: Proceedings of the 22nd International Conference on Data Engineering (2006)
12. Kyoung-Sook, K., Si-Wan, K., Tae-Wan, K., Ki-Joune, L.: Fast indexing and updating method for moving objects on road networks. In: Proceedings of the 4th International Conference on Web Information Systems Engineering Workshops (2003)
13. Liu, F., Hua, K.A.: Moving query monitoring in spatial network environments. Mob. Netw. Appl. (2012)

14. Hadjieleftheriou, M., Kollios, G., Gunopulos, D., Tsotras, V.J.: On-line discovery of dense areas in spatio-temporal databases. In: Hadzilacos, T., Manolopoulos, Y., Roddick, J., Theodoridis, Y. (eds.) SSTD 2003. LNCS, vol. 2750, pp. 306–324. Springer, Heidelberg (2003)
15. mobiThinking (2013)
16. Mouratidis, K., Papadias, D., Bakiras, S., Tao, Y.: A threshold-based algorithm for continuous monitoring of k nearest neighbors. IEEE Trans. Knowl. Data Eng. **17**, 1451–1464 (2005)
17. Mouratidis, K., Papadias, D., Hadjieleftheriou, M.: Conceptual partitioning: an efficient method for continuous nearest neighbor monitoring. In: Proceedings of the ACM SIGMOD International Conference on Management of Data (2005)
18. Mouratidis, K., Yiu, L., Papadias, D., Mamoulis, N.: Continuous nearest neighbor monitoring in road networks. In: Proceedings of the 32nd International Conference on Very Large Data Bases (2006)
19. Nehme, R.V., Rundensteiner, E.A.: SCUBA: scalable cluster-based algorithm for evaluating continuous spatio-temporal queries on moving objects. In: Ioannidis, Y., Scholl, M.H., Schmidt, J.W., Matthes, F., Hatzopoulos, M., Böhm, K., Kemper, A., Grust, T., Böhm, C. (eds.) EDBT 2006. LNCS, vol. 3896, pp. 1001–1019. Springer, Heidelberg (2006)
20. Ni, J., Ravishankar, C.V.: Pointwise-dense region queries in spatio-temporal databases. In: IEEE 23rd International Conference on Data Engineering (2007)
21. Nutanong, S., Tanin, E., Shao, J., Zhang, R., Kotagiri, R.: Continuous detour queries in spatial networks. IEEE Trans. Knowl. Data Eng. **24**, 1201–1215 (2012)
22. Prabhakar, S., Xia, Y., Kalashnikov, D.V., Aref, W.G., Hambrusch, S.E.: Query indexing and velocity constrained indexing: scalable techniques for continuous queries on moving objects. IEEE Trans. Comput. **51**, 1124–1140 (2002)
23. Tao, Y., Papadias, D., Sun, J.: The TPR*-tree: an optimized spatio-temporal access method for predictive queries. In: Proceedings of the 29th International Conference on Very Large Data Bases, VLDB 2003, vol. 29, pp. 790–801. VLDB Endowment (2003)
24. Šaltenis, S., Jensen, C.S., Leutenegger, S.T., Lopez, M.A.: Indexing the positions of continuously moving objects. SIGMOD Rec. **29**(2), 331–342 (2000)
25. Wang, H., Zimmermann, R.: Processing of continuous location-based range queries on moving objects in road networks. IEEE Trans. Knowl. Data Eng. **23**, 1065–1078 (2011)
26. Wen, J., Meng, X., Hao, X., Xu, J.: An Efficient approach for continuous density queries. Front. Comput. Sci. **6**, 581–595 (2012)
27. Xia, T., Zhang, D.: Continuous reverse nearest neighbor monitoring. In: Proceedings of the 22nd International Conference on Data Engineering (2006)
28. Xiong, X., Mokbel, M.F., Aref, W.G.: SEA-CNN: scalable processing of continuous K-Nearest neighbor queries in spatio-temporal databases. In: Proceedings of the 21st International Conference on Data Engineering (2005)
29. Yiu, M.L., Tao, Y., Mamoulis, N.: The Bdual-tree: indexing moving objects by space filling curves in the dual space. Very Large Data Bases J. **17**, 379–400 (2008)

Query Operators for Comparing Uncertain Graphs

Denis Dimitrov[✉], Lisa Singh, and Janet Mann

Georgetown University, Washington, DC 20057, USA
dd322@georgetown.edu

Abstract. Extending graph models to incorporate uncertainty is important for many applications, including citation networks, disease transmission networks, social networks, and observational networks. These networks may have existence probabilities associated with nodes or edges, as well as probabilities associated with attribute values of nodes or edges. Comparison of graphs and subgraphs is challenging without probabilities. When considering uncertainty of different graph elements and attributes, traditional graph operators and semantics are insufficient. In this paper, we present a prototype SQL-like graph query language that focuses on operators for querying and comparing uncertain graphs and subgraphs. Two interesting operators include ego neighborhood similarity and semantic path similarity. Similarity operators are particularly useful for comparison queries, the focus of this paper. After motivating and describing our operators, we present an implementation of a query engine that uses this query language. This implementation combines a layered and service-oriented architecture and is designed to be extensible, so that simple operators can be used as building blocks for more complex ones. We demonstrate the utility of our query language and operators for analyzing uncertain graphs based on two real world networks, a dolphin observation network and a citation network. Finally, we conduct a performance evaluation of some of the more complex operators, illustrating the viability of these operators for analysis of larger graphs.

Keywords: Graph query language · Comparison queries · Similarity queries · Uncertain graphs

1 Introduction

Graphs and networks have become a ubiquitous type of data. Traditional graph models contain nodes, edges, and their respective attributes. Figure 1(a) shows a small example containing two nodes (Kate and Joe), one edge (the solid line between the nodes), and attributes associated with the nodes (*node_id* and *gender*). It is assumed that nodes and edges in the graph exist, and attributes have a known value or a null value. However, many data sets contain uncertainty about vertex existence, edge existence, and attribute values. Figure 1(b) shows an example where existence probabilities and attribute value confidences have

© Springer-Verlag Berlin Heidelberg 2015
A. Hameurlain et al. (Eds.): TLDKS XVIII, LNCS 8980, pp. 115–152, 2015.
DOI: 10.1007/978-3-662-46485-4_5

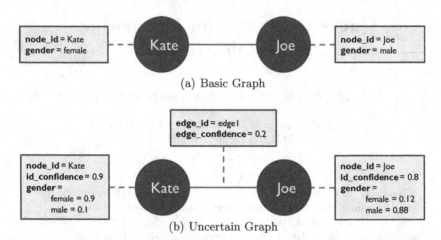

Fig. 1. Two different representations of graph data. 1(a) shows the representation for traditional graph data. 1(b) shows an example of an uncertain graph.

been added. Data from many different application domains that can be represented using uncertain graph models include disease transmission networks with disease transmission probabilities, observed terrorist networks with node existence probabilities, and physical computer networks with associated reliability probabilities.

While basic queries involving most probable attribute values and the identification of nodes with high certainty can be handled by relational, graph, and probabilistic query engines, none of the corresponding database systems have query languages that focus on (or in most cases even handle) operators specifically designed for comparison of uncertain graphs. In this paper, we are interested in introducing operators that are relevant to graph comparison and to uncertainty. While many operators are important for graph queries or probabilistic queries, they have already been introduced in previous literature and are, therefore, not the focus of this paper. Interest in uncertain graph analysis is emerging, but is still in its infancy [22,27,37,38].

Our motivation for uncertain graph comparison arises from two completely different motivating scenarios - uncertainty occurring during scientific observation and uncertainty resulting from data analysis. We now describe the importance of uncertain graph comparison in each of these examples.

Observational scientific data: Observational scientists study animal societies in their natural settings, often, with the purpose of understanding the social relationships and behaviors within the society [20]. Such social network data can be captured as a graph, where nodes represent observed animals and edges between nodes represent sightings of both animals together. A researcher observing a particular animal may be uncertain about its identification, relationships, features, or behavior. This uncertainty can be expressed as existence probabilities between 0 and 1, associated with nodes and/or edges, and attribute value confidences, represented as discrete probability distributions over the set of possible categorical attribute values. Analyzing and comparing uncertain graphs can be useful for answering questions about similarities between local neighborhoods of different animals, changes in animal sociality over time, diffusion of behaviors, differences among animal subgroups across locations, and observation bias across researchers, to name a few.

Analysis output data: A second setting we consider involves machine learning algorithms generating uncertain graphs that can be used as the basis for prediction, generalization, and statistical analysis. One specific example is a *node labeling* algorithm. Node labeling algorithms attempt to predict the label (attribute value) of nodes in a graph [28]. For example, they can be used to predict the topics of each publication in a citation network, or to predict which customers will recommend products to their friends using a customer network, or to predict the political affiliations of people in a social network [29]. The input to a node labeling algorithm is a partially observed graph. The output of such algorithms is an uncertain graph containing a probability distribution across the possible set of labels or attribute values for each node. Comparing and contrasting these uncertain graphs to each other or to a ground-truth graph allows researchers to analyze the performance of different machine learning algorithms, experiment with a single algorithm under different assumptions, and examine the graph dataset by highlighting parts of data where the algorithms disagree in their predictions or perform poorly.

Figure 2 shows a small node labeling example. The graph on the left side is a ground truth graph containing the true labels of the *sex* attribute for each node. The node labeling task is to correctly label node A. In other words, assume

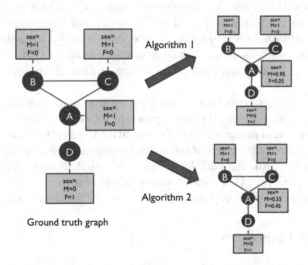

Fig. 2. A node labeling example. Different node labeling algorithms generate an uncertain graph containing a prediction of the *sex* attribute.

node A does not have a known label and use different algorithms to predict the label. In this example, two algorithms run and each gives its prediction for the *sex* attribute of node A. Algorithm 1 is more confident about predicting 'Male' than algorithm 2, but both agree on the label. Having operators that compare these uncertain values for different attributes and components of the graph is important when analyzing the output of different algorithms.

Contributions: This paper is an extended version of [11]. In [11], we introduced the following. (1) A basic SQL-like language, which incorporates uncertain graph analysis and comparison operators, while taking advantage of existing SQL capabilities. The semantics of this proposed language is a combination between relational database and uncertain graph semantics, part of which is novel and part of which is necessary for the language to be applicable. (2) A set of composable, comparative operators for uncertain graphs, where the previous literature focuses on single graph operators, on specific graph algorithms in the presence of uncertainty, or on operators for multiple certain graphs. We introduce operators for estimating similarity between graphs, nodes, edges, and their attributes, including finding a common subgraph that exists across two graphs containing edges with high certainty and identifying a set of nodes that have the same predicted node label across two uncertain graphs. (3) A novel system framework that uses a combination of a layered and a service oriented architecture, and is extensible, modular, and expandable, allowing for easy integration of new operators. The novelty of our design is the focus on extensibility and modularity. Traditionally, databases construct query trees whose set of possible operations is predefined. This design allows for query optimization by applying a set of rewrite rules. Our approach provides a flexible mapping of operators to their implementation. The query engine can, therefore, support easy integration of new operators without affecting the existing ones and without requiring significant changes to the framework itself. While we do not claim this design to be better than traditional query processing, we show it to be a viable alternative with advantages when new operators are being designed. (4) An initial implementation of our query framework and a demonstration of our approach for two case studies.

Along with those contributions, this extended version also includes the following additional contributions. (1) New operators related to directed graphs that are particularly useful in the context of observational scientific data. (2) A significantly more detailed discussion of the query language, the query engine, and the system architecture. (3) An additional case study that focuses on the new hierarchical path operator. (4) A performance analysis of some of our more complex operators that includes a comparison to a traditional relational query engine.

2   Related Literature

Storage, analysis, and manipulation of graph data is a vast area of interest for both the research community and industry. In particular, there are multiple graph databases in existence: [2,4–7], to name a few, offering efficient data storage and access, as well as scalability and transaction management. The query options sometimes include proprietary APIs (Neo4j Traverser) or proprietary SQL-like query languages [6]; in other cases ([1,2,5–7]) there is support for Gremlin [3] or SPARQL [25]. SPARQL is a query language for the RDF format with similarities to our approach in terms of semantics and SQL-like syntax, including joins and the capability to retrieve and combine data from several graphs. In comparison to our language, SPARQL offers more flexible pattern matching, but is more restricted in that its standard data types are XML-based, its set of operators is more limited, and it does not offer extended SQL-like constructs such as MERGE BY and SPLIT BY presented in Sect. 4. Gremlin is a relatively simple but powerful language for graph traversal and manipulation, supporting built-in functions such as union, difference, intersection that are applicable to graph comparison. Neither of these languages focuses on uncertain graphs. Similarly, many of the languages suggested in existing research [8,9,13–15] often do not consider uncertainty during graph analysis and comparison across multiple graphs.

Querying similar graphs in graph databases has been studied in recent years [40]; however, existing works mainly focus on structural information and connectivity. Uncertainty is often incorporated in the context of specific algorithms [16,17,23,24,30,37,38,42]. These problems are important in answering some of the possible uncertain graph queries, yet our goal is to create a more comprehensive set of albeit simpler uncertain comparison operators. Other researchers study uncertainty arising from approximate queries rather than uncertain data [39]. Moustafa et al. [22] propose a graph model for reasoning about different types of uncertainty that arise in different real world entities and relationships that can be represented in graphs. They introduce the probabilistic entity graph (PEG) and then propose algorithms for subgraph pattern matching. Our comparative operators can be used with PEG graphs or other variations that model uncertainty within a graph [27].

Probabilistic databases, on the other hand, typically support queries based on the concept of Possible World Semantics [18,32,35]. Recently researchers have extended the Possible World Semantics to uncertain graphs [27,36,41]. While applicable for many problems, this concept is different from our focus on graph comparison regardless of the nature of the underlying probabilities. We do, however, build upon operators for comparison of probabilistic attributes [32], as they are applicable for uncertain graph attributes.

Finally, there are a number of visual graph tools, including [10,29]. Excellent for visual comparison of uncertain graphs, they could complement rather than substitute the capability to execute user-defined queries.

We pause to mention that while we implemented our query language independent of the SQL query language or other ones that have been proposed in the graph and probabilistic databases literature, we could have chosen to build it above existing query languages. As shown in the performance analysis section, it is feasible for many of our operators. Our choice to not do so resulted from our interest in designing a language and its constructs in a way that allows for easy manipulation and comparison of uncertain graphs. Incorporating either uncertainty or graph constructs into existing query languages is cumbersome at best.

Those database query languages were not designed to handle comparison queries for uncertain graphs. Further, because we are interested in continually adding more operators, we also preferred a design approach that was particularly extensible.

3 Probabilistic Formulation

Throughout the subsequent sections we use the following background definitions, underlying assumptions, and notation. As mentioned in Sect. 1, the object of interest is an uncertain graph. It is a generalization of a deterministic graph, incorporating uncertainty about vertex/edge existence and attribute values. We now formally describe the elements of the graphs being studied.

Uncertain Graph. An uncertain graph $G = (V, E, A^V, A^E, PA^V, PA^E)$ has a non-empty finite set of vertices, $V = \{v_1, \ldots v_m\}$, and a finite set of undirected edges, $E = \{e_1, \ldots e_n\}$, where each edge e_y is a pair of vertices, $e_y \in V \times V$, and $V \times V = \{(v_i, v_j) | v_i \in V, v_j \in V\}$; $A^V = \{A_1, \ldots A_p\}$ is a set of (certain) attributes for vertices; $A^E = \{A_1, \ldots A_q\}$ is a set of (certain) attributes for edges; $PA^V = \{PA_1, PA_2, \ldots PA_r\}$ is a set of uncertain attributes for vertices; and $PA^E = \{PA_1, PA_2, \ldots PA_t\}$ is a set of uncertain attributes for edges. The attributes are consistent across vertices and across edges respectively, i.e. all vertices have the same schema and so do all edges. We refer to both edges and vertices as *graph elements*.

Certain Attributes. The set of all certain attributes is defined as $A = \{A_1, A_2, \ldots A_s\} = A^V \cup A^E$. Given an attribute $A_j \in A^V$, its domain D_j, and a vertex $v_i \in V$, we associate a value $b_k \in D_j$ with the pair (v_i, A_j) and denote it using the notation $a(v_i, A_j) = b_k$. Every vertex has an identifying attribute. We refer to this attribute as the *node_id*, or *id* for shorthand. Similarly, an edge attribute value is denoted as $a(e_i, A_j) = b_k$. When we are generically speaking about an attribute value on either a vertex or edge graph element, as shorthand, we will use a_{ij}.

Structural Uncertainty. To express structural uncertainty, we store our confidence about existence of the corresponding graph element as one of the attributes in A^V and A^E: $\exists A_j : a_{ij} \in [0, 1], \forall i \in [1, m]$ and $\exists A_j : a_{ij} \in [0, 1], \forall i \in [1, n]$. Henceforth, we refer to this attribute as '$conf$' or '$confidence$'.

Uncertain Attributes. By analogy, the set of all uncertain attributes is $PA = \{PA_1, PA_2, \ldots PA_o\} = PA^V \cup PA^E$. Uncertain attributes allow the data model to express semantic uncertainty in the graph. The value of an uncertain attribute PA_j is a set of pairs of each possible attribute value and a probability associated with each possible value. For example, an uncertain attribute *sex* with value domain $\{male, female\}$ reflects the researcher's uncertainty about the sex of the observed animal. For a specific vertex, the set of its value pairs could be $\{(male, 0.8), (female, 0.2)\}$.

More precisely, value domain VD_j is the constrained (discrete) domain of possible values associated with attribute PA_j. The value domain is ordered and we use the notation a_j^t to designate the t-th member of VD_j, where $t \in [1, |VD_j|]$. Continuing with the previous example, $a_j^1 = male$, $a_j^2 = female$.

PD_j is the domain of uncertain attribute PA_j: $PD_j = \{\{(a_j^t, f(a_j^t) : \forall a_j^t \in VD_j\}$: for all probability distribution functions $f(x)$ over the value domain $VD_j\}$. In other words, the domain PD_j is the infinite set of all permissible values for uncertain attribute PA_j, where each of these values corresponds to a different possible probability distribution function and thus in itself represents a set of pairs of each possible value from the value domain VD_j and the corresponding pdf output.

Given an uncertain attribute $PA_j \in PA^V$ and a vertex $v_i \in V$, we associate a value $c_k \in PD_j$ with the pair (v_i, PA_j) and denote it using the notation $pa(v_i, PA_j) = c_k$. Similarly, given an attribute $PA_j \in PA^E$ and an edge $e_i \in E$, we associate a value $c_k \in PD_j$ with the pair (e_i, PA_j) and denote it using the notation $pa(e_i, PA_j) = c_k$. As shorthand, when the type of attribute (vertex vs. edge) is not significant, we use pa_{ij}.

By analogy to using a_j^t to refer to members of value domain VD_j, the shorthand p_{ij}^t refers to the corresponding probability $f(a_j^t)$, associated with value a_j^t for vertex v_i. We define the set of uncertain attributes for a particular vertex v_i as $PA(v_i) = \{PA_j : PA_j \in PA^V \text{ and } pa(v_i, PA_j) \neq null\}$. Similarly, the set of uncertain attributes for a particular edge e_i is defined as $PA(e_i) = \{PA_j : PA_j \in PA^E \text{ and } pa(e_i, PA_j) \neq null\}$.

According to these definitions, our data model supports uncertain attributes only with discrete probability distribution. Future work will consider extending this model to support continuous uncertain attributes.

Assumptions. We make the following general assumptions about the uncertainty in the graph and the form of comparison:

Assumption 1: The existence probability of an edge is assumed to be conditional upon the existence of its endpoints.

Assumption 2: Uncertain attributes contain probabilities associated with each possible value from their domain, expressing the likelihood that the attribute takes on this particular value.

Assumption 3: We make no assumptions about the nature of probability values assigned to the graphs that we need to compare. In other words, the analyst can decide if the probabilities are marginal or posterior.

Assumption 4: When comparing two graphs $g1$ and $g2$, we assume that the following partial mapping exists between their elements: (1) the vertex mapping consists of a bijective mapping function for those vertices that are mapped, plus a set of unmapped vertices in each of the graphs $g1$ and $g2$; and (2) edge mapping is equivalent to vertex mapping, with the added constraint that edges $g1.e$ and $g2.f$ can be mapped to each other only if both of their endpoint vertices are also mapped. We refer to two graphs with this property as *aligned graphs*.

Assumption 5: Alignment is assumed to be based on the id of the element: elements from graph $g1$ are mapped to elements with the same unique id in graph $g2$; they are unmapped if there is no corresponding element with the same id.

4   Query Language

In this work we create a new query language that incorporates necessary operators for uncertain graph comparison. While we create a new language, it makes sense to leverage people's SQL knowledge and use the SQL semantics to handle graphs, graph elements, and attributes as relations when possible. In other words, we will use the notion of a relation, but we will allow a tuple to contain any graph element, e.g. a vertex, an attribute value, or a collection of any graph elements, including an entire graph. As we will show, the ability of a graph to be a value in a tuple of a relation is important for graph comparison.

We chose to base our query language on SQL because it is a mature, proven, and well-known language. While we do not claim that it is the best language for the purpose, we believe it is sufficient for expressing a wide range of uncertain graph comparison queries using our set of operators. Using these operators directly with traditional SQL was certainly another option; however, there are several disadvantages. On syntactic and semantic level, a dedicated query language allows the flexibility for any modifications that best suit the specifics of uncertain graph analysis. On implementation level, extending an existing SQL query engine effectively would mean using a relational database as storage for uncertain graph data. While this is reasonable, a number of interesting graph and probabilistic databases have arisen. Therefore, we wanted to use an approach where the query language was not restricted to a particular storage type. Having the ability to query any underlying database in a single consistent manner was an important goal. Our query language accomplishes this by leveraging important SQL semantics without tying the language to relational databases.

Operations from SQL. Our query language supports the major SQL operations, such as *SELECT, FROM, JOIN, WHERE, GROUP BY, HAVING*, and *ORDER BY*, introducing modifications and extensions to accommodate the specifics of graph comparison. For example, the *FROM* operation can extract individual nodes, edges, both nodes and edges, attributes from the specified graph (creating a tuple for each of them), or return the graph as a whole in a relation as a single tuple. An example that returns a table containing edges e from graph $g1$ with high existence probability ($conf(e) > 0.5$) along with the confidence of existence ($conf(e)$) sorted by the edge confidence is as follows:

```
SELECT e, conf(e)
FROM g1 TYPE edge AS e
WHERE conf(e) > 0.5
ORDER BY conf(e) DESC
```

New Operations. We introduce two new operations to support the manipulation of collections. These operations are *MERGE BY* and *SPLIT BY*. To describe these operations, suppose that α is a collection containing a set of values, for example, the set of values for a categorical attribute.

The *SPLIT BY* operation is used to separate each element in the collection α into separate tuples. In other words, in the result relation of the SPLIT BY operation the original tuple is replaced with a set of tuples, one for each element in α. The remaining columns in each of these tuples are unchanged. Intuitively, this operation is used to "flatten out" or "unnest" a relation when α contains a set of values instead of a single value.

Suppose we have the following query:

```
SELECT n, location
FROM g1 TYPE node AS n
SPLIT BY mpv(n.location) AS location
```

In this query, $g1$ is the graph of interest and n is a column of type vertex. *location* is an uncertain attribute containing all the possible values for location for each vertex in n, and the $mpv()$ operator gets one or possibly several most probable values from the attribute. In this example, the SPLIT BY operation transforms the single row, collection result obtained from the $mpv()$ operator to a multi-row result. In other words, it extracts values (in this case, the most probable values) from a collection (in this case, the *location* attribute) into separate tuples. This query produces the result shown in Fig. 3.

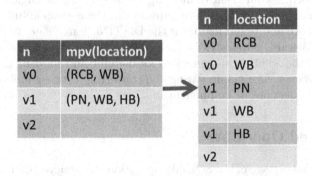

Fig. 3. Illustration of simple SPLIT BY operation.

The MERGE {DISTINCT} (BY | ALL) operation is the semantic opposite of the SPLIT BY operation. Similar to GROUP BY, the MERGE BY operation returns a relation that contains one tuple for each distinct column value referenced in the clause, or distinct combination of values in case of multiple columns. Unlike GROUP BY, the MERGE BY operation retains the original values in each of the remaining columns by merging them into a collection - one collection for each resulting tuple and column. Therefore, all columns remain visible to subsequent non-aggregate operators. For illustration, assuming that vertices have an attribute *size* equal to one of the "small", "medium", and "large", the following example query will produce the result in Fig. 4. In this example, vertices from graph $g1$ form column n, whose type is vertex. The vertices with the same attribute value can be merged into a single collection using the MERGE BY operation.

```
SELECT n, size
FROM g1 TYPE node AS n
MERGE BY n.size AS size
```

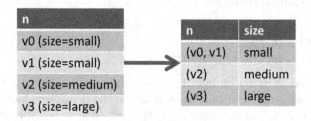

Fig. 4. Illustration of simple MERGE BY operation.

The MERGE BY operation can be applied not only to columns, but to an arbitrary mix of columns and expressions. Omitting them altogether (specified as MERGE ALL), is equivalent to merging a relation r into a single tuple, where each column contains a collection of values from the corresponding column for all tuples of r. An additional feature is the DISTINCT modifier. When enabled, the operation discards duplicates for each column that is turned into a collection. As illustrated, the addition of operations for merging and splitting elements in a relation are necessary for manipulating and converting between individual graph elements and collections of graph elements.

5   Proposed Operators

While we take advantage of the SQL-like operations to retrieve, filter, sort, group, and join data, individual operators are used within each of these clauses to specify the required behavior. As shown in the query example, the same operator can be re-used with several operations, subject to rules between aggregate vs. non-aggregate operators and operations.

In order to effectively query uncertain graphs, we need operators for attributes, graph elements, local subgraphs and graphs, and hierarchical elements. While this leads to a large number of operators, many of which are quite intuitive and straightforward, we also introduce some novel operators as well. Therefore, this section begins by highlighting important operators for each of the mentioned targets (attributes, graph elements, local subgraphs, and hierarchical elements).

This is followed by a detailed discussion of similarity operators, since they are most relevant to the uncertain graph comparison tasks outlined in the motivating examples and span some of the previous categories of operators.

The different operator examples are based on the sample graph in Fig. 5. In this example, there are two nodes, $v1$ and $v2$, and one edge $e1$. Each node has three attributes, id, $conf$, and loc, where id is the unique identifier, $conf$ is the existence probability, and loc is an uncertain attribute that contains 5 possible values (PN, WB, HB, EA, RCB). This uncertain attribute represents the location where different dolphins are observed. Finally, the edge has two attributes, id and $conf$. Parallel to their node counterparts, id is the unique identifier and $conf$ is the existence probability of the edge.

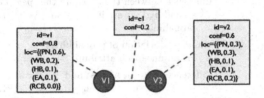

Fig. 5. Sample network for operator examples.

5.1   Operators for Uncertain Attributes

Operators in this category can be used to answer queries about values and probabilities associated with one or more uncertain attributes. Because the attribute value represents a discrete probability distribution, the proposed *attribute operators'* functionality ranges from simply extracting the probability that an attribute has a specific value ($valueCertainty()$) to the most/least probable value ($mpv()$ and $lpv()$) to analyzing the shape of the distribution ($peakToAvgDist()$). Table 1 lists operators related to uncertain attributes. The leftmost column of the table contains the name of the operator. The middle column describes the operator and the rightmost column shows an example of the operator in use. For example, the $peakToNextDist()$ operator is used to identify uncertain attributes with a dominant (peak) probability that significantly exceeds the probabilities for the remaining values.

5.2   Operators for Graph Elements

Graph element operators may be incorporated into a query, such as the one in the introductory example in Sect. 4, to identify strong/weak connections within a single graph ($conf()$) or to compare the confidence of the corresponding elements across two aligned graphs ($sim()$), isolating the elements not only based on their low or high confidence, but also on whether the two graphs agree or differ significantly ($compBin()$). Table 2 lists the operators for graph elements.

Table 1. Uncertain Attribute Operators: These operators answer queries about values and probabilities associated with one or more uncertain attributes.

Operator	Description	Example
$mpv()$, $lpv()$	most/least probable attribute value	$mpv(v1.loc) =$ "PN"
$valueCertainty()$	probability that an attribute has a specific value from the domain	$valueCertainty(v1.loc,$ "WB") $= 0.2$
$maxValueCertainty()$, $minValueCertainty()$, $avgValueCertainty()$, $medianValueCertainty()$	max, min, mean, or median probability among all probabilities associated with an attribute	$maxValueCertainty$ $(v1.loc) = 0.6$
$peakToAvgDist()$	difference between the max and the average certainty	$peakToAvgDist$ $(v1.loc) = 0.4$
$peakToNextDist()$	difference between the max and the second-highest attribute value probability	$peakToNextDist$ $(v1.loc) = 0.4$
$valueCertaintyDev()$	standard deviation of probabilities for an uncertain attribute	
$valueCertaintyRange()$	difference between highest and lowest probability of an attribute	$valueCertaintyRange$ $(v1.loc) = 0.6$
$sim()$	similarity score between two uncertain attributes of the same type, typically in the range [0, 1]. It is generally measured between the two sets of their respective attribute values and probabilities	The specific similarity measures are described later in this section

Similarly to Table 1, this table has three columns for the operator name, description, and an example using the operator. Some of these operators are designed to be used together. For example, the $bin()$ operator returns a *true* or *false* value. Given a threshold, the operator returns a *true* when the confidence of the graph element is higher than the threshold and a *false* when it is lower than the threshold. The confidences of two graph elements can then be compared using $compBin()$ operator. It uses the output of the $bin()$ operator to determine the probabilistic relationship between the two graph elements, returning a 'high', 'low', or 'opposite' value. A value of 'high' indicates that both graph elements have a high confidence. A value of 'low' indicates the contrary. A value of 'opposite' indicates that the two elements have divergent confidences.

5.3 Operators for Local Subgraphs and Graphs

Sometimes, when graphs are analyzed, the comparison of interest is not on a graph element. Instead, it is on a part of the graph containing multiple vertices and multiple edges. One common level of analysis is the ego network level. An ego network of a vertex v is the subgraph that contains v, its neighbors, and

Table 2. Graph Element Operators: Most of these operators serve to query and analyze the confidence of existence of a single graph element or relative to another vertex or edge. Other operators in this group aggregate the results from attribute-level operators for the given graph element.

Operator	Description	Example
$conf()$	confidence of element's existence	$conf(v1) = 0.8$
$bin()$	true or false bin, corresponding to high or low $conf()$ relative to a threshold	$bin(v1, 0.5) = $ true
$compBin()$	"high", "opposite", or "low", depending on the relationship between the output of the bin() operator applied to each of the two operands	$compBin(v1, v2) = high$
$magnitudeDiff()$	difference between confidence of existence of 2 elements	$magnitudeDiff$ $(v1, v2) = 0.2$
$diffSignificance()$	whether the absolute value of magnitude difference is above a threshold	$diffSignificance$ $(v1, v2, 0.1) = true$
$valueCertaintyScore()$	average $maxValueCertainty()$ of all uncertain attributes of the element	$valueCertaintyScore$ $(v1) = 0.6$
$sim()$	similarity score between 2 elements of the same type (vertices or edges), typically in the range $[0, 1]$	

the edges connecting v to its neighbors [34]. Common usage examples of *graph* and *ego-net operators* are provided in Table 3. Operators allowing for structural graph comparison based on graph alignment include: *intersect(g1, g2)*, *union(g1, g2)*, *difference(g1, g2)*, and *bidirectionalDifference(g1, g2)*. For example, by intersecting the ego networks of two specific dolphins in the same graph, the analyst can discover their common friends. The graph reconstruction operator, *toGraph()*, can be used in a query to derive a subgraph based on specified conditions. For example, to obtain a subgraph of high-confidence elements, it can be combined with the *bin()* operator and MERGE BY clause. While this operator is not as sophisticated as pattern matching [15], it does provide the capability for subgraph filtering based on a flexible set of conditions.

5.4 Operators for Trees

We also introduce operators for capturing hierarchical relationships and tree structures in the graph. For example, some networks contain family relationships such as parent/child. Identifying matrilines within the graph using tree-related

Table 3. Graph and ego-net operators: These operators can be used for structural and semantic comparison of subgraphs and graphs.

Operator	Description
$egoNet()$	given a vertex v_i, returns the set of vertices and edges that are part of v_i's ego-network, including v_i itself
$egoSim(v_1, v_2)$	similarity score between two ego-networks defined by their center vertices v_1 and v_2, respectively, typically in the range $[0, 1]$
$intersect()$, $union()$, $difference()$, $bidirectionalDifference()$	creates a new graph that represents, respectively, an intersection, union, difference, and bi-directional difference of two graphs
$toGraph()$	recreates a graph from a set of vertices and edges
$toElements()$	breaks down a given graph into a set of vertices and edges

operators makes it possible to see which attributes or behaviors are most probable within a family and which are not. Table 4 contains a list of operators related to hierarchical analysis. In addition to several traditional tree operators such as $treeDepth()$, $hasChildren()$ and $children()$, this category includes operators allowing for subtree comparison that can be useful when trying to understand properties of social networks.

As an example, suppose a node attribute indicates lineage in the network, perhaps parent-child relationships. This data can be useful for understanding when different behaviors occur within a family. Is the behavior consistent within single branches of the family tree or do some of the nodes in the branch exhibit the behavior while others do not? To help with this, we introduce the $switchRatio$ operator, described in Sect. 5.6. Finally, the $countFeature$ operator supplements the different flavors of the $switch$ operator by giving the number of nodes possessing a given attribute value, although its functionality in most cases could be replicated using other operators at the expense of query complexity and readability. Its application is not restricted to trees, as it can be used on any collection of elements.

5.5 Similarity Operators

Uncertain Attribute Similarity. The $sim()$ operator is one of our novel operators for comparing uncertain attributes pa_{ij} and pa_{lj}. In the proposed set of measures, similarity is classified as either structural or semantic. The former identifies the similarity between the general shapes of the two distributions, ignoring the attribute values and their arrangement relative to each other. For example, attribute $\{(a, 0.8), (b, 0.1), (c, 0.1)\}$ should be considered structurally equivalent to attribute $\{(a, 0.1), (b, 0.1), (c, 0.8)\}$, as both have a dominant value

Table 4. Subtree operators: Necessary for understanding hierarchical components of the graph.

Operator	Description
roots()	returns a set of vertices, representing root nodes for every tree in the given graph, i.e. all vertices having child nodes but not parent nodes. Optionally, vertices without child nodes may be included
hasParents(), *hasChildren*()	test for existence of parent and child nodes of a given vertex, respectively
parents(), *children*(), *siblings*()	return a set containing, respectively, the parent nodes, child nodes, or siblings of a given vertex
tree()	extracts from a graph the tree rooted at the given node by creating a set of all descendant nodes and the corresponding directed edges that connect these nodes to the tree
treeDepth(), *treeSize*()	depth and size (number of nodes), respectively, of the tree rooted at the given node
switchRatio(), *switchRatioPositive*(), *switchRatioNegative*()	a score in the range $[0, 1]$ measuring variation of the given attribute value among the descendants of the given node. Defined as the number of edges in the hierarchy where an attribute value of the node changes in the specified way between the parent and child divided by the total number of edges in the hierarchy
countFeature()	count of elements in the given set with a particular attribute value

(peak) of 0.8. Structural similarity is useful for discovering certain envelope patterns. For example, a pattern with a single dominant value would suggest that observers (or algorithms) were able to establish the value with a higher degree of certainty than if the pattern is flat - regardless of which exactly value is dominant. We support two structural similarity measures, entropy ratio and absolute distance ratio, where the entropy ratio compares the distribution spread for the specified uncertain attribute and the absolute distance ratio compares the magnitude of the distance between the different uncertain attribute values. For example, the absolute distance ratio equals $\frac{AD(pa_{ij})}{AD(pa_{lj})}$, where absolute distance is calculated as $AD(pa_{ij}) = \sum_{t=2}^{|VD_j|} |p_{ij}^t - p_{ij}^{t-1}|$ and by analogy, for pa_{lj}. To correctly reflect structural similarity through absolute distance, probability sets in both attributes must first be sorted.

The semantic similarity, on the other hand, compares probabilities between the corresponding attribute values. An instance of an uncertain attribute can be represented as a histogram. We refer to each possible attribute value as a 'bin' in the histogram, conceptually containing the associated probability. This representation allows us to use a number of measures that have been proposed for histogram similarity. They generally fall into two categories - bin-by-bin and cross-bin approaches [26]. The *bin-by-bin* similarity compares the contents of only corresponding bins, or in our case, probabilities for the same attribute values in two attribute instances. *Cross-bin* measures, on the other hand, compare non-corresponding bins. This is possible only if the *ground distance* between pairs of non-corresponding attribute values is known. In this work we focus on the following bin-by-bin similarity measures, because they are useful across many domains. Instead of selecting one, we implement the ones most frequently used in the literature [26]:

1. Default: $sim(pa_{ij}, pa_{lj}) = 1 - \frac{\sum_{t=1}^{|VD_j|} |p_{ij}^t - p_{lj}^t|}{2}$
2. Minkowski-Form Distance
3. Histogram intersection
4. K-L divergence

We refer you to [26] for details about each of these measures.

Ego Network Similarity. The *egoSim*() operator uses a variety of similarity measures and algorithms depending on user-specified constraints and on ego network containment within the same or different graphs. For measuring similarity between two ego networks (or ego-nets), the two center nodes are mapped to each other, each of the non-center nodes from the first subgraph is mapped to 0 or 1 non-center nodes from the second subgraph, and vice versa. For ego-net similarity, we assume that multiple edges between a pair of vertices are not allowed.

We now intuitively describe different types of ego-net similarity. They are the cornerstone of our uncertain comparative operators, allowing researchers to better compare graph substructures, not just entire graphs or single graph elements. The different cases of ego-network similarity are outlined in Fig. 6. We consider all the permutations, since different application domains may be interested in different forms of similarity. Depending on alignment between the two ego-nets, similarity can be aligned and unaligned. In the aligned case, the mapping is determined by the alignment scheme. If no alignment scheme is chosen (not aligned case), the elements are mapped between the two ego-networks in a way that maximizes similarity.

Ego-net similarity can be structural, semantic or both. Structural similarity only takes into account the existence or confidence of existence of vertices and edges in each mapped pair between the two ego-nets, while ignoring attributes and their values. Structural similarity is subdivided into topological, probabilistic-topological, and comparison count. The user can select the similarity measure that is most applicable to the comparison.

Before we define the various similarity measures, we need to introduce some additional notation. Let $o1$ and $o2$ be the center vertices of the two ego-networks,

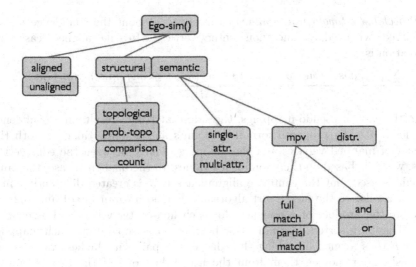

Fig. 6. Classification of ego-network similarity.

$eg1$ and $eg2$, respectively. Let $|eg1|$ and $|eg2|$ be the size of the corresponding ego-net, measured as number of non-center vertices, i.e. the degree of the center vertex. In the aligned case, let a mapping rm be expressed as (e, v, e', v'). In this notation, e and v are an edge and non-center vertex from $eg1$, while e' and v' are the corresponding mapped edge and non-center vertex from $eg2$. Let the alignment set RM be the set of all mappings between vertices and edges from $eg1$ and $eg2$, excluding the center vertex. By definition, RM does not include unmapped edge-vertex pairs. To represent this case, we define a pseudo-mapping pm in the same way as the regular mapping rm, except that one side of the pseudo-mapping is always null: $pm = (e, v, e', v')$, $(e = null, v = null)$ *or* $(e' = null, v' = null)$. Let the set PM contain all pseudo-mappings pm between $eg1$ and $eg2$. Then, the complete alignment set M is defined as $RM \cup PM$. $|RM|$, $|PM|$, and $|M|$ designate the sizes of the corresponding sets. In a similar fashion, we define a pair of mapped uncertain attributes as $mp = (ua, ua')$, where $ua = pa(v, PA_j)$ and $ua' = (v', PA_j)$ share the same definition (domain and order of values). The set MP contains all pairs of mapped attributes between a pair of mapped vertices. Note that one side of $mp \in MP$ can be *null*.

Topological similarity compares the structure of the two ego-nets based on the existence of their elements, but not on confidence values associated with existence. It is calculated as follows:

- In the aligned case: $\dfrac{|RM|}{max(|eg1|, |eg2|)}$
- In the unaligned case: $\dfrac{|eg1|}{|eg2|}$, if $|eg1| < |eg2|$, else $\dfrac{|eg2|}{|eg1|}$

Probabilistic-topological similarity takes into account the confidence values associated with edges and non-center vertices. In the aligned case the calculation is:

$$1 - \frac{\sum_{m \in M} abs\_value(conf(m.e) * conf(m.v) - conf(m.e') * conf(m.v'))}{|M|}.$$

In the case of pseudo-mappings, the side that is null results in a confidence product of 0. The proposed formula accounts at the same time for both the number of mapped vertices and the mapping quality between the edge-vertex pairs, which is based on their confidence values. In the unaligned case, the same formula is used, but the complete alignment set M is created differently - in a manner similar to the merge-sort algorithm. For both ego-nets $eg1$ and $eg2$, we calculate the product of confidences for each non-center vertex and associated edge. These are sorted descending for both ego-nets separately. Each mapping rm in RM is formed by taking the edge-vertex pair with highest values from each list and removing them from the list. When one of the lists is empty, pseudo-mappings pm are formed using the remaining elements of the other list. This algorithm, as most of our algorithms used in the unaligned case, is an approximation.

Comparison count is simply a count of aligned non-center nodes between the two ego-networks. It is useful when the researcher is interested in an absolute similarity measure, related to the size of the ego-networks, rather than in a ratio between 0 and 1 that is returned by the topological and probabilistic-topological similarity. In the aligned case, the value is $|RM|$. In the unaligned case, the value is: $min(|eg1|, |eg2|)$.

Semantic similarity, on the other hand, ignores confidence and derives the similarity score by only using similarity measures between the individual nodes and edges in the mapped pairs. In the aligned case, similarity is measured by aggregating similarities between pairs of attributes with the same name and definition, belonging to each pair of aligned vertices. In the unaligned case, alignment is chosen in an attempt to maximize the total similarity of all attribute pairs between aligned vertices - usually by greedy heuristics.

Depending on the number of attributes under consideration, the measure can be either single- or multiple-attribute. In both of those cases, similarity between a pair of uncertain attributes can be estimated using different measures. We propose two of them: mpv and distribution similarity. In the aligned case, attribute similarity is always calculated as:

$$\frac{\sum_{rm \in RM} similarity\_measure(rm.v, rm.v')}{|RM|}$$

where $similarity\_measure(rm.v, rm.v')$ varies based on one of the single attribute or multiple attribute cases.

Case 1 - Single attribute. Let PA_j be the selected uncertain attribute, and $ua = pa(v, PA_j)$, $ua' = pa(v', PA_j)$. Then one of the following measures can be used for similarity:

- MPV. The user can select between a partial or a full match between the attribute's sets of mpv values:
  $similarity\_measure(rm.v, rm.v') = 1$, if $mpv(ua) \cap mpv(ua') \neq \emptyset$, $else\, 0$, or
  $similarity\_measure(rm.v, rm.v') = 1$, if $mpv(ua) \equiv mpv(ua')$, $else\, 0$. The result is also 0 if $ua = null$, $ua' = null$, or both.
- Distribution. In this case, the distribution of certainty values for an attribute is used:
  $similarity\_measure(rm.v, rm.v') = sim(ua, ua')$, where sim is any user-selected semantic uncertain attribute similarity measure.

Case 2 - Multiple attributes. More options and complexity exist when considering multiple attributes. One of the following measures can be used for similarity:

- MPV. Between each pair of mapped uncertain attributes, we use the same similarity measure as in the case of single attribute: $smpv(ua, ua') = 1$, if $mpv(ua) \cap mpv(ua') \neq \emptyset$, $else\, 0$, or $smpv(ua, ua') = 1$, if $mpv(ua) \equiv mpv(ua')$, $else\, 0$. Because we deal with multiple mapped attributes $mp \in MP$, the total $similarity\_measure$ for the set MP can be derived in different ways from the similarity measure $smpv$ between each mp pair.
  - 'AND' - the result is 1 if $smpv(mp.ua, mp.ua') = 1$, $\forall mp \in MP$, $else\, 0$.
  - 'OR' - the result is the average of pairwise attribute similarity for all mapped attribute pairs: $\frac{\sum_{mp \in MP} smpv(mp.ua, mp.ua')}{|MP|}$
- Distribution. Between each pair of mapped uncertain attributes, we use the same similarity measure as in the case of single attribute: $sdistr(ua, ua') = sim(ua, ua')$, where sim is any user-selected semantic uncertain attribute similarity measure. Because the values returned by $sdistr$ are not restricted to either 1 or 0, we do not apply the 'AND' case in dealing with multiple mapped attributes $mp \in MP$. The result is derived by averaging the pairwise attribute similarity for all mapped attribute pairs: $\frac{\sum_{mp \in MP} sdistr(mp.ua, mp.ua')}{|MP|}$.

In the unaligned case, we restrict the similarity measure to a single attribute for considerations of computational complexity. Even in the case of a single attribute, the brute force approach for finding the alignment that would maximize similarity is highly inefficient in some cases. In those cases, we propose using a greedy heuristics, similar to the merge-sort algorithm, that reduces running time but does not guarantee optimality. As in the aligned case, the user has a choice of two similarity measures, MPV and distribution based similarity.

5.6   Tree Branch Attribute Similarity

In many graph data sets, hierarchies exist within the graph structure. For example, in a social network, node attributes may indicate lineage such as parent-child relationships. When exploring data, observational scientists are sometimes interested in understanding when different behaviors occur within a family. Is the behavior consistent within single branches of the family tree or do some of the nodes in the branch exhibit the behavior while others do not? To help with this, we introduce the *switchRatio* operator. The *switchRatio* operator is the number of edges in the hierarchy where an attribute value of the node changes between the parent and child divided by the total number of edges in the tree. The *positiveSwitchRatio* is the number of edges in the hierarchy where the attribute value changes from not existing in the parent (FALSE) to existing in the child (TRUE) divided by the total number of edges. *negativeSwitchRatio* is the opposite - the attribute value exists in the parent, but not in the child. A *switchRatio* of 0 indicates homogeneity of the behavior in the tree branch. A higher *switchRatio* indicates more variation between parent-child nodes in the tree. A *switchRatio* of 1 indicates that every child has a different value for the attribute than its parent. This operator is an important supplement to a simple count of the number of nodes in the graph with a particular attribute value. It gives the analysts and scientists some initial insight into how the behavior is diffusing throughout the graph.

5.7   Other Operators

In addition to operators related to uncertain graph comparison, the proposed query language supports general operators, most of which are commonly present in many other languages, including SQL, e.g. aggregate operators, logical operators, set operators, etc.

5.8   Route Operators

While path operators are central to graph query languages, their use is not as central as similarity for uncertain graph comparison. Some operators that are useful in this context include: comparing high confidence path existence between two nodes or ego-nets, comparing high confidence shortest paths, and comparing connected components when taking into account the confidence of existence of graph elements. While we have not implemented them in our query language, we consider them useful for uncertain graph comparison and leave them for future work.

6   High Level System Architecture

This section describes the high-level architecture of the query engine. Broadly, we combine concepts from service-oriented design and a layered system architecture to create a highly extensible framework. The remainder of this section begins by describing our design priorities, followed by a high level explanation of the architecture itself and a discussion of the query compilation and operator composition.

6.1   System Goals

Our highest priority design goals in developing the query engine architecture and prototype implementation include:

Extensibility. Because we intend to continue building upon our initial query language, allowing for extensibility at all levels was our highest priority. At the lowest level, the system must allow easy integration of any additional operators and operations. When concepts that do not fit in the existing implementation are introduced, for example, aggregate operators, it is desirable to minimize the required changes to the query processing framework. We refer to this as mid-level extensibility. At a high level, the design must provide room for new system capabilities, such as plugging in different data storage implementations.

Operator composition. Operators sometimes re-use the functionality of other existing operators. For example, vertex and ego-network similarity operators build upon different attribute similarity operators. Because we anticipate that being a common situation, the system should provide re-use of existing operators to the programmer, who creates new operators. The user can also compose operators implicitly within the limits of the query language by creating expressions or within the limits of the pre-programmed sub-operator selections, such as choosing an underlying attribute similarity measure.

Adaptability. Capabilities to introduce future optimizations specific to our data model and query language without restricting the implementation to a particular platform or data storage.

6.2 System Overview

To meet these goals, we use a combination of layered and service-oriented architecture, illustrated in Fig. 7. The main component of this architecture is the query engine, a lightweight and generic platform for deployment of modules responsible for the individual steps in the query processing workflow, such as

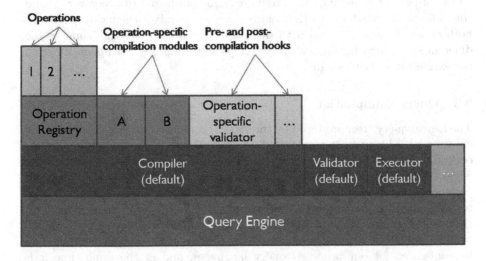

Fig. 7. Different components of the layered architecture.

parsing, compilation, optimization, validation, and execution. The set of included modules, represented as services, is not pre-defined, making it different from traditional database query engines.

The engine offers two important capabilities: service configuration and service lookup. The former allows parameter tuning without code recompilation, including deploying the same implementation under different configurations. The latter allows flexible and dynamic linking of services, e.g. transparent replacement of the underlying data storage implementation. The individual modules are designed with the goal of decoupling them from each other and, in turn, they can be customized by plugging in implementations of their sub-components. For example, an operation registry is the sub-component that provides the default mapping between operators and their compiled representations. For integrating simple operators, it is sufficient to add a reference to the registry.

The query execution process is as follows. It begins when the Parser module transforms the textual representation of a query into an abstract syntax tree (AST), which we refer to as *logical query*. The Compiler module translates the AST using a post-order traversal of operations in the logical tree into an internal representation suitable for optimization and execution. Next, the Optimizer module generates and evaluates several alternative execution plans, choosing the best one. The Executor is the key module where the operations and operators that make up the query are executed. The Validator module can be invoked at different stages to ensure compliance with the pre-defined rules. The Facade and Connector modules provide the interface for interaction between external systems and the query processing workflow. Data Store serves to retrieve the data requested in the query and convert the graph(s) into an incidence list based internal representation, which includes vertex objects and edge objects. Each vertex object has an instance variable pointing to a collection object that lists the neighboring edge objects. In turn, each edge object points to the two vertex objects at its endpoints. Attribute maps are linked to every vertex and edge object and conform to the common vertex or edge schema, respectively, both of which are associated with the graph. We now go through some of the different steps, focusing on how one develops and composes new operators given the extensible system design.

6.3 Query Compilation

The logical query tree, produced by the Parser, conforms to a simple and generic data model. The query is represented as a logical tree, composed of operations, constants, and variables, which in turn can contain expression subtrees. Tree edges represent data flow in the direction of the root. The Compiler performs post-order traversal of operations of the logical query tree, mapping logical nodes to their compiled counterparts. The mapping between operations is given with the operation registry. For more customization, compiler modules can be assigned to operations that require special handling. This way, the default compiler can be configured for our proposed query languages, and at the same time it is generic enough to be used with other query languages conforming to the same logical structure. Adding new operations to the compiler in most cases requires minimum effort, involving only registering them with the operation registry.

In our design, we maintain a separate structure for a *compiled query*, which is executed by the query executor module. Mapping a logical query to the compiled query is the task of the compiler module. Decoupling the two structures from each other provides a clean separation between the specification of the user request and the internal operations taking place during query execution, such as optimizations. An additional benefit is that the queries can be pre-compiled and parameterized at a later time using the *logical variable* component of the logical query. This separation does not enforce pre-compilation. If interpretation is more desirable, compilation can be done at run-time.

6.4   Developing and Composing Operators

Traditionally, databases construct query trees whose set of possible operations is predefined. This design allows for query optimization by applying a set of rewrite rules. Our approach differs because we provides a flexible mapping of operators to their implementation. This allows the query engine to support easy integration of new operators without affecting the existing ones and without requiring significant changes to the framework itself.

Developing an operator involves implementing a simple interface with two methods. The first method allows the Executor to set the operator's input parameters. Then, the second method is called, in which the operator performs its calculations over these supplied parameters and returns the result. In the simple case, no other code is required. Adding the operator to the configuration of the OperationRegistry is sufficient to incorporate it into the query language, as the registry is used for both compilation and execution.

Composing an operator using operators that are already in the language is also straightforward, as the framework supports their lookup and execution from the dependent operator. For example, ego-network semantic similarity re-uses one of the existing attribute level similarity operators, as selected by the user, to derive similarity for the ego-network as a whole. During execution, each operator has access to the Context, from which it can retrieve its configuration and any other data previously bound to the Context. Because the user does not directly specify the nested operators, it is not possible to pass their configuration as regular parameters; instead, the configuration is bound to the enclosing operator and passed down to the nested operators, which use the Context mechanism to retrieve these parameters.

The Context also provides implementation instance of a simple invocation interface, which abstracts calling other existing operators explicitly from within an operator and decouples their implementations. In addition to this static operator composition, for even more flexibility, when the nested operator is not known until run-time, an operator can define an arbitrary logical query, compile, and execute it dynamically.

7   Detailed Operator Use Cases

To show the utility and composition ability of our operators, we have integrated our query engine with Invenio [12,31], a visual analytic tool for graph mining. Our query engine and the Invenio tool are both written in Java. The application analyzes graphs in main memory and visualizes different projections of them. Using the two motivating scenarios presented in Sect. 1, we highlight a subset of our operators in three different case studies, two that utilize a dolphin association network and one that uses a citation network.

7.1   Queries to Support Ego-Net Analysis

The Shark Bay Research Project studies dolphins in Shark Bay, Australia for over 30 years [21]. Our data set includes demographic data about approximately 800 dolphins, represented as graph nodes with certain attributes (id, conf, dolphin name, birth_date) and uncertain attributes (sex_code, location, mortality_status_code). Survey data about social interactions between these dolphins are captured as approximately 29,000 edges with attributes (id, conf).

Our team met with researchers on the Shark Bay Research Project and developed a list of typical queries that observational scientists would like the capability to issue when analyzing this dolphin social network and its inherent uncertainty:

- Selecting the number of associates and sex composition of associates for male and female dolphins, respectively, using the most probable value of the *sex_code* attribute.
- Visualizing the union, intersection, difference, and bi-directional difference between the ego-networks of a particular dolphin during two different years, where the confidence of relationship existence is above a specified threshold.
- Finding the common associates (friends) of two specific dolphins with a relationship confidence above a certain threshold.
- Finding all dolphins having associates whose most probable location is different from their own.
- Calculating a measure of structural and semantic similarity between ego-networks of two particular dolphins.
- Selecting the subgraph that consists only of dolphins linked by observations with low confidence of existence (lower than a specified threshold). The results of this query tell researchers if observers are having difficulty identifying certain dolphins.

The query in Table 5 is an example that shows counts by sex of dolphins seen together (task 1). The inner query selects pairs of dolphins seen together and uses *SPLIT BY* to split into a set of rows the collection that is returned by the *adjacentVertices*() operator. The outer select produces counts for each possible sex combination, grouping the nodes based on the most probable sex_code. Researchers can use the resulting table to see that dolphins who are most probably males are seen together more often than any of the other combinations.

Table 5. Sample query and its result: counts by sex of dolphins seen together.

MALE	MALE	9930
MALE	FEMALE	6184
FEMALE	MALE	6184
FEMALE	FEMALE	6092

```
SELECT sex, sexAdj, count(adj) AS cntFriends
FROM (
    SELECT *
    FROM
(
    SELECT n, adjacentVertices(n) AS adj
    FROM g1 TYPE node AS n
)
SPLIT BY adj
)
GROUP BY first(mpv(n.sex_code)) AS sex,
  first(mpv(adj.sex_code)) AS sexAdj
```

The second task focuses on determining the union, intersection, difference, and bi-directional difference between the ego-networks of a particular dolphin during two different years. It introduces a time component. The results for a particular dolphin are displayed in graph format in Fig. 8. It is easy to see that

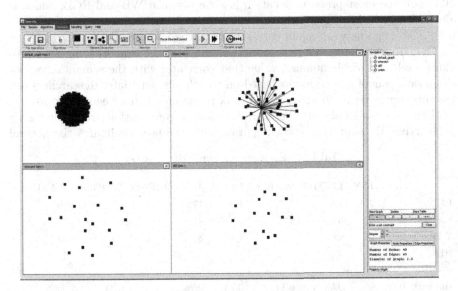

Fig. 8. Clockwise from upper left: complete dolphin network, union, intersection, difference of ego-networks of dolphin 'JOY' between years 2010 & 2009.

the dolphin has almost as many new associates as repeat associates, i.e. occurring during both years. Researchers can then visually explore who these associates are, what sex they are, etc., to gain more insight about dolphin sociality.

To validate the significance of our similarity operators, we evaluate one of the more complex measures. We estimate the ego-network semantic similarity between dolphin *JOY* and other dolphins in the same graph, in absence of alignment, using the most probable value of the location attribute. By picking dolphins with different characteristics, we can demonstrate the behavior and validity of the chosen similarity measure. For example, we discovered that the average ego-net similarity by location is twice as high for dolphins located in the same primary area as JOY, e.g. RCB: 0.28 vs 0.14. This is the expected result, since dolphins are likely to have associates mostly in their primary location.

To compare uncertain ego-nets, we randomly chose several dolphins from different locations with high and low similarity relative to *JOY's*. For every dolphin under consideration, we ran a query to retrieve their most probable location, their ego-network location similarity to that of *JOY*, and a breakdown by location of the dolphin's ego-network. The results are summarized in Table 6. They are consistent across the two cases of same and different location. Both *LITTLE* and *WHELK* differ from *JOY* in their most probable location; however, the ego-network's location composition between *LITTLE* and *JOY* results in a much higher similarity score. *PUCK* and *JOYSFRIEND* reside in the same location as *JOY*, share many associates with her, and have a very similar distribution of associates by location. These commonalities lead to a particularly high similarity score. *MYRTLE*, on the other hand, who only shares 88 out of 147 associates with *JOY* despite the same location, is average in similarity. For *WANDA*, the most probable location is a tie between WB and RCB, which is also reflected in having associates from mostly those locations. This difference with *JOY's* ego-network again corresponds to the lower similarity.

Overall, examining different cases confirms that the similarity measure provides a relevant single numeric value that correlates with the semantic composition of a pair of ego-networks based on the chosen attribute. Researchers can use this simple result to identify and rank potentially similar ego-networks.

The query also shows operator re-use and composition from the user's perspective. By supplying context parameters, the user configures the general

Table 6. Ego-network similarity results.

	JOY	LITTLE	WHELK	PUCK	JOYSFRIEND	MYRTLE	WANDA
RCB	211	125	1	171	172	92	56
EA	38	5	43	32	47	6	7
WB	28	48		7	8	45	33
HB		4				1	4
PN						5	
sim with JOY		0.56	0.14	0.75	0.78	0.44	0.32
primary loc	RCB	WB	EA	RCB	RCB	RCB	WB, RCB

ego-net similarity operator. Specifying the mpv-based similarity measure and attribute name causes the similarity operator to re-use the $mpv()$ operator to retrieve the most probable attribute value.

7.2   Queries to Support Node Labeling Algorithm Comparison

In the second scenario, we examine the output of two different node labeling algorithms. For this analysis, we use the CiteSeer paper citation data set from [28]. It consists of 3312 scientific publications classified into one of six topics. In the citation network each publication is a node and each citation is an edge. We use partially observed citation data to predict the probability distribution of the topic attribute of each paper by applying two different classification algorithms. The queries of interest deal with understanding the similarities and differences between most probable node labels across the two classification algorithms and include:

– Selecting the papers, whose topic certainty is significantly higher in one uncertain graph when compared to the other.
– Selecting the papers, for which the predicted discrete probability distribution differs the most between the two graphs, using different attribute similarity measures, e.g. KL divergence, Minkowski-form distance, and histogram intersection.
– Counting the number of papers that are misclassified by both models.
– Selecting the papers, which are misclassified with high confidence by both classifiers.

The example query below retrieves the count of papers misclassified by one of the models with confidence over 0.75. The inner query joins nodes from the predicted graph *ica* and the ground truth graph *gt* based on their id. The outer query filters these tuples using nested operators to express the desired criteria and selects the count of remaining tuples.

```
SELECT count(g1Node) as cntMisclassified
FROM (
    SELECT g1Node
    FROM ica TYPE node AS g1Node
    JOIN
    SELECT gtNode
    FROM gt TYPE node AS gtNode
    ON g1Node.id == gtNode.id
)
WHERE and(
    greaterThan(maxValueCertainty(g1Node.label), 0.75),
    isEmpty( setIntersect(mpv(g1Node.label), mpv(gtNode.label)) )
)
```

Some of the results we found using this data set are as follows: model 1 misclassified fewer documents (83) than model 2 (103); of the documents misclassified by both classifiers (65), both models misclassify them with the same label; and 8 of the 10 largest ego networks were in the area of information retrieval.

7.3 Queries to Support Analysis Related to Diffusion of Behaviors

The first two use cases demonstrated how our language can be used to answer a wide range of questions about a dolphin social network in the presence of observational uncertainty and about a citation network with uncertainty introduced by node labeling algorithms. In this final use case, we focus on queries that support investigation of diffusion of behaviors. Observational scientists are interested in understanding the diffusion of different foraging behaviors, exhibited by dolphins in Shark Bay. To support this type of analysis, we consider useful queries related to two particular foraging behaviors - *sponging* and *snacking*. Sponging refers to a foraging behavior where dolphins find and wear marine sponges on their beaks to help them flesh out hiding prey on the seafloor in deep channels. The sponges help protect the dolphins from sharp rocks or shells when searching for these buried fish [19,33]. Researchers in Shark Bay have learned that sponging is transmitted vertically from mother to child. Snacking is a foraging behavior in which a dolphin swims belly up and chases small fish, trapping them at the water surface. Researchers in Shark Bay are uncertain about how this behavior is transmitted. In this use case, we use operators to help identify any matrilineal relationships between the foraging behaviors and the dolphins.

In our dataset, sponging and snacking behaviors are two different binary node attributes. The certain attribute values are set to 'true' if a dolphin exhibits the corresponding behavior. The dataset additionally includes 544 directed edges designating a relationship between a mother and a child dolphin. Using the directed graph operators described earlier, we develop simple hierarchical queries and establish that the graph contains 192 matrilineal trees and 383 single nodes that do not have any incoming or outgoing directed edges. The trees have maximum depth of 3 and maximum size of 20 nodes, with an average size of 3.8.

To either hypothesize about behavior diffusion or confirm a hypothesis, researchers can use the proposed operators to study where the behavior occurs: across parent-child relationships (or generations) throughout the matrilineal tree, within the dolphin's ego-network, or within the dolphin's siblings.

Using our *count* operator and *GROUP BY* operation, we identify the following about the dolphins in our dataset:

- Out of 1119 dolphins in this dataset, only 44 dolphins use the sponging foraging behavior while 110 use snacking. This represents less than 4 % and 10 % of the dolphin population, respectively.
- Only 1 dolphin exhibits both sponging and snacking.
- The proportion of males and females exhibiting each behavior varies. Fewer males sponge than females (9 males vs 33 females), but more snack than females (59 males vs 43 females)[1].

---

[1] The numbers do not add up to the total count of sponging and snacking dolphins, because the sex for some of these dolphins cannot be established with certainty.

We now use the hierarchical operators to try to find support for matrilineal behavior diffusion throughout the graph. Using a group-by query on mother-child pairs, we see that all sponging child dolphins have a mother that sponges. This supports the idea that the behavior may be transmitted vertically. We also see that the reverse is not true: a sponging mother has children who do not sponge. Both of these results are consistent with previous research findings by Shark Bay researchers.

In contrast, children who exhibit the snacking behavior do not always have mothers who exhibit the behavior (38 snacking moms vs 33 non-snacking moms). The reverse is also similar. A snacking mother has a similar number of children who use the snacking foraging tactic as do not (38 snacking vs 44 non-snacking). In addition, non-snackers generally have non-snacking mothers by a large margin - 429 vs 44. As determined in response to the following sample query returning a breakdown by mother snacking, child sex_code, and child snacking, snacking is spread evenly across sex groups for every combination of mother-child snacking behavior. Consequently, in this dataset, it appears that snacking is not a foraging behavior that is always transmitted vertically.

```
SELECT motherSnack, childSexCode, childSnack, count(n) AS cnt
FROM (
    SELECT *
    FROM (
     SELECT n, children(n) AS child
     FROM gt TYPE node AS n
    )
    SPLIT BY child
)
GROUP BY n.snacking AS motherSnack, child.sex_code AS childSexCode,
 child.snacking AS childSnack
```

As a final interesting note, there is only one mother who has children that use both sponging and snacking. This is interesting because there is little overlap between the subpopulations that exhibit each of these behaviors.

Queries dealing with behavior within siblings indicate that both sponging and snacking are independent of the corresponding behavior of a dolphin's siblings, i.e. all spongers have between 0 and 6 siblings total and all snackers have between 0 and 7 siblings total, with any number of siblings among them exhibiting the sponging or snacking behavior, respectively. The only consistent observation is that for every sponging dolphin who has siblings, at least one of the siblings is a non-sponger.

The following query demonstrates how counting sponging siblings and total siblings of each dolphin is a straightforward task, facilitated by the *countFeature* along with *siblings* operator.

```
SELECT
 countFeature(siblings(n), "sponging") AS cntSponging,
 size(siblings(n)) AS cntSiblings
FROM gt TYPE node AS n
WHERE equals(n.sponging, "true")
```

We now analyze the family trees using the *switchRatio* operator. For spong-ing, the *positiveSwitchRatio* = 0, reconfirming that all spongers have a sponging mother. Many trees, including the three largest trees, have a *switchRatio* = 0, i.e. completely homogeneous. All of the remaining trees except one have *switchRatio* = *negativeSwitchRatio* ≥ 0.5. In other words, there are many children who have sponging mothers yet do not sponge themselves.

When querying for snacking behavior, we discover that many trees have a *switchRatio* = 0, because they do not contain any snackers. Within the remaining trees of depth 1 or 2, the *switchRatio* varies from 0.14 to 1.0 depend-ing on the total number of nodes. Considering that snackers are in minor-ity, the lower number usually occurs when the root is a non-snacking dolphin, and therefore, the *positiveSwitchRatio* > 0. A higher *switchRatio*, accom-panied by a *negativeSwitchRatio* > 0, usually occurs when the root node is a snacking dolphin. The more children the root dolphin has, the larger the *negativeSwitchRatio*. For example, the trees rooted at KWI and QUO respec-tively, have the following numbers (Table 7):

Table 7. Switch ratio for two example trees.

dolphin name	*swtichRatio*	*positiveSwitchRatio*	*negativeSwitchRatio*
KWI	0.33	0.33	0
QUO	0.4	0	0.4

These trees have approximately the same number of nodes, structure, depth, and same number of snackers (see Fig. 9). However, snackers are related in a different way to other snackers. Hence the *switchRatio*'s are equal, but the positive and negative ones are different.

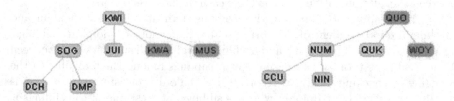

Fig. 9. Trees rooted at nodes KWI and QUO respectively. Green-colored nodes repre-sent snacking dolphins (Color figure online).

Due to the large size of ego-networks (on average, 94 associates for a sponging dolphin and 164 associates for a snacking dolphin), it is not surprising that the networks tend to include a mix of spongers, snackers, and those dolphins not exhibiting either behavior. After querying the minimum, maximum, and average number of associates for each sponger and snacker, we find that both for sponging and snacking dolphins the average ratio of spongers and snackers to the total size of the ego-network (0.23 and 0.36, respectively) is significantly higher than the graph average of approximately 0.04 and 0.1, respectively.

These simple queries on directed graphs and trees can help observational scientists begin hypothesizing about the impact of the networks on the diffusion of behaviors. This use case is also a testimony to the extensibility of the query language design, as new operators were added specifically for the analytical task of studying behavior diffusion without impacting existing operators.

These three example cases demonstrate how our language can be used to enable scientists to formulate a wide range of ad-hoc queries that analyze and compare uncertain graphs and hierarchies without the need for custom programming.

8   Performance Evaluation

To illustrate the viability of our proposed query language, this section presents a performance evaluation on synthetic graphs ranging from 100 s of nodes to 1 million nodes. Similar to many real world, non-synthetic networks, the graphs we study are sparse in terms of number of edges. We begin by comparing our uncertain graph query language to the relational query language. Then for some interesting operators that do not translate easily into traditional SQL or procedural SQL, we present the runtime query performance for queries involving those operators, again, highlighting the viability of the language.

We pause to mention that none of the operators or queries containing the operators have been optimized. This un-optimized query evaluation is presented to illustrate the following: (1) the reasonable scalability of the operators; (2) the types of operators that are not suited for a standard relational query language; (3) future directions that can be explored to improve the performance of the query language while maintain its strength for extensibility.

8.1   Performance Comparison to Standard Relational Query Language

Our query language was developed in the Java programming language. Because the proposed language is similar to SQL, we compare the performance of our query engine against an implementation based on a relational database. To that end, we created a fully functional implementation of a subset of our operators in PostgreSQL 9.3. The graphs and their elements are stored in tables shown in the schema in Fig. 10. Operators are implemented through user-defined functions in SQL and PL/pgSQL, using similar algorithms as in our Java operators.

For this evaluation, we execute equivalent queries on the same data sets in both implementations. We generate three synthetic data sets of varying size: 10,000 nodes and 100,000 edges; 100,000 nodes and 1,000,000 edges; and 1,000,000 nodes and 1,000,000 edges. Each of these graphs has a mix of randomly generated string,

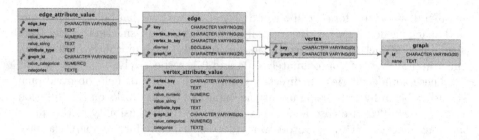

Fig. 10. PostgreSQL schema for performance evaluation

numeric, and categorical attributes associated with their vertices, two attributes of each type. Categorical attributes have up to seven distinct attribute values.

All experiments were run on a MacBook Pro with 16 GB of RAM, 2.6 GHz quad core Intel i7 processor, and 512 GB SSD drive. We average the execution time of 5 runs, after discarding the first run. Removing the first run allows us to ignore the variations due to query parsing, loading data into buffers, etc.

There are several major implementation differences that we want to point out. Our Java query engine reads the entire graph(s) into in-memory graph data structures, before executing the query. Because we want the ability to tie this language to different data storages, we chose this approach. For this reason, we do not implement indices. RDBMS that pre-index their primary keys will, by default, have an advantage over our unindexed implementation. In other words, if the RDBMS only needs a subset of the data, i.e. it does not need a full table scan, it will access the data using an index and also take advantage of internal cache buffering. To mitigate the I/O costs, our implementation can keep the data in memory and re-use it for all subsequent queries, until space is needed for retrieving a different graph.Therefore, we provide performance results both including and excluding the time required for initial graph loading.

The second important consideration is the inherent disadvantage of storing graphs in a relational database, which results in multiple joins for queries requiring graph traversal. Our approach of abstracting graph structures from relational tables and processing on them directly is expected to be more efficient, especially for path and graph traversal algorithms. Therefore, when choosing our queries, we begin with queries that do not require graph traversal beyond one hop. In the last subsection, we consider a query that involves longer path traversal.

Query 1: The first query returns the set of all nodes, for which a particular categorical attribute has a maximum value certainty above a threshold. The results are summarized in Table 8. We see that PostgresSQL performs better, by using primary key indices. In our case, we are reading the entire data set in memory. On the other hand, if the graph is cached in memory, the subsequent executions exhibit comparable performance with PostgreSQL.

Table 8. Running times of filtering by maxValueCertainty query (msec).

Graph size (nodes + edges)	Java - total time	Java - query exec. only	PostgreSQL time
10 000 + 100 000	977	259	95
100 000 + 1 000 000	13 375	670	1 438
1 000 000 + 1 000 000	47 810	13 716	11 061

Query 2: This query returns the difference between ego-networks of two particular vertices, selected by id. With slight modifications (i.e. difference instead of intersection), this query is borrowed from one of our earlier use case examples, where we were interested in finding common friends of two dolphins. The query was chosen to compare performance for cases when the RDBMS optimizer cannot take advantage of indices. The SQL execution plan employs a nested loop instead. Table 9 shows that the running times of our Java implementation are very close to the running times of the previous query and again, dominated by the I/O cost. The suboptimal PostgreSQL execution is not only much slower than the previous query, but is also worse than a full table scan based execution in our engine.

Table 9. Running times of ego-network difference query (msec).

Graph size (nodes + edges)	Java-total time	Java-query exec. only	PostgreSQL time
10 000 + 100 000	1 055	350	4 828
100 000 + 1 000 000	13 719	650	61 835
1 000 000 + 1 000 000	49 614	15 899	77 310

8.2   Operator Scalability Evaluation

Some of our operators are not easily implemented in PostgresSQL, for example, operators involving attribute similarity. Therefore, in this subsection, we demonstrate the scalability of our query language implementation for a more complex query without a direct comparison with relational databases.

Query 3: Our third example returns the average semantic attribute similarity between the ego-network of a particular vertex and the ego-networks of all other vertices in the graph that satisfy a condition based on one of the uncertain categorical attributes. The query includes where clauses, a join, and aggregation, in addition to the relatively expensive similarity operator applied to a large number of vertices. The actual query follows. For this query, $n1$ and $n2$ are vertices in graph gt.

```
{egosim_attr_name="cat0", egosim_align="false",
 egosim_measure="MPVAttributeSimilarity"}
SELECT average(semanticSimilarity(n1, n2)) AS avgEgoSim
FROM (
SELECT n1
FROM gt TYPE node AS n1
WHERE n1.id = ?
JOIN
SELECT n2
FROM gt TYPE node AS n2
)
WHERE in("a", mpv(n2.cat0))
```

The results on the three synthetic data sets are shown in Table 10. For these sample graphs, the execution time scales linearly with the number of nodes.

Table 10. Running times of ego-network similarity query (msec).

Graph size (nodes + edges)	Java-total time	Java-query execution only
10 000 + 100 000	1 348	640
100 000 + 1 000 000	17 097	5070
1 000 000 + 1 000 000	61 799	26 010

8.3 Path Traversal Query Evaluation

We now consider a query related to the new switch operator. This query involves path traversals since we are considering hierarchies in the graph. As mentioned earlier, queries with underlying path traversals are generally slow in a relational database since multiple joins are required over the RDBMS schema.

Query 4: This query simply calculates the *switchCount* of a single tree. To abstract the processing time from disk I/O, we experiment with the same graph containing 10 000 vertices and 100 000 edges, and instead choose trees of varying sizes. Also, in this data set, trees represent the subgraph of a directed acyclic graph, rooted at a designated vertex. Consequently, there may be multiple paths through some of the vertices and hence, we measure the tree size in edges rather than nodes. As expected, Table 11 shows that our implementation scales very well to path queries, while the relational implementation struggles with larger trees.

8.4 Discussion

The experiments confirm that the performance bottleneck in our engine is the I/O cost of loading the entire graph in memory. Profiling showed that reading

Table 11. Running times of switch count query (msec).

Tree size as number of edges	Java-total time	Java-query exec. only	PostgreSQL time
17	980	260	461
1 036	1 017	282	27 585
9 103	1 055	339	241 492

and handling attributes has a major performance impact, which at this stage we mitigated by loading a graph in parallel using several threads. Depending on the number of attributes, we realized performance gains ranging from insignificant to up to 3–3.5 times compared to single-threaded loading. A further improvement would be loading attributes lazily on demand, only if they are used in a query. It should be noted that the RDBMS implementation is also directly affected by the number of attributes, because the cardinality of attribute tables equals the number of attributes × the number of vertices and edges, respectively. A more sweeping optimization and one of our planned future directions involves retrieving only the required data, either introducing indices or completely integrating with an external graph database.

Even without such optimizations, the current implementation is suitable for practical application with graphs on a scale of a million vertices and edges. Running several additional queries adapted from our use cases confirmed that the query engine scales well to the relatively larger synthetic graphs, using commodity hardware. The observed memory and execution time at different graph sizes gives confidence that through vertical scalability, the engine should be capable of processing larger graphs using a multi-core/high memory infrastructure such as Amazon Cloud Services.

Despite not being built with performance as a priority, for some queries our implementation provides comparable performance to an RDBMS-based implementation that takes advantage of a number of optimizations already built into the database. For other queries, such as the ones including operators that require graph traversal, our engine outperforms the relational-based storage. In fact, the performance of our implementation seems comparatively more stable and predictable across different queries we tested, while our query language still benefits from the same SQL semantics. In addition, our engine performance improves dramatically when ignoring I/O costs, i.e. if usage patterns permit running different queries on the same graphs that are already loaded into memory.

Finally, while implementations of some operators in traditional SQL are concise and efficient, we found that other operators and queries are non-trivial, fully benefiting from the additional semantics of our proposed language. For example, using graphs as first-class citizens allows for convenient representation of entities such as ego-networks and other subgraphs. For some operators, such as semantic similarity, we did not see an efficient and straightforward implementation option using only SQL and PL/ pgSQL.

9   Conclusions and Future Directions

Graphs have become a ubiquitous data model with application in multiple domains. For many applications, it is important to account for uncertainty, either inherent or introduced during data analysis by extending the traditional graph model. We address the need to analyze and compare graphs and subgraphs by considering the graph structure, the graph semantics, embedded hierarchical structures, all in the context of uncertainty. In this paper, we proposed a SQL-type language with a set of composable comparative operators for uncertain graphs and their elements. Our language takes advantage of developers' knowledge of SQL by incorporating the generic logic of existing SQL and extending it to consider graphs and uncertainty. We developed a query engine implementation using an extensible, modular system framework that combines layered and service-oriented architecture. The novelty of this approach lies in the focus on breadth of operator creation and composition, as opposed to an in-depth focus on the optimization of a single operator. We then presented case studies that demonstrated the utility of the proposed language and operators for analyzing different aspects of real-world uncertain graph datasets. Finally, we presented a simple performance evaluation and comparison to a relational database implementation that confirms the viability of our approach for relatively large graphs.

There are several future directions to consider. First, because our focus was on operation creation and composition, we have not developed a query optimizer. We need to investigate generic and specific query optimization for our operators. If we focus on optimizing base operators that are used in the composition of a number of different operators, then many different types of queries will benefit from the optimizations. Adding indexing is also important. In general, our initial prototype implementation can benefit in terms of performance from a more flexible subgraph retrieval, indexing, and optimizations on certain similarity measures. Finally, there are other operators that could be beneficial for uncertain graph comparison including additional path and routing operators, graph mining algorithms, e.g. community detection, and notions of uncertain time-varying graphs.

Acknowledgments. This work was supported in part by the National Science Foundation Grant Nbrs. 0941487 and 0937070, and the Office of Naval Research Grant Nbr. 10230702.

References

1. ArangoDB graph database. http://www.arangodb.org/
2. DEX graph database. http://www.sparsity-technologies.com/dex
3. Gremlin language for graph traversal and manipulation. https://github.com/tinkerpop/gremlin/wiki
4. Neo4j graph database. http://neo4j.org/
5. Oracle spatial and graph option. http://www.oracle.com/technetwork/database-options/spatialandgraph/overview/index.html

6. OrientDB document-graph DBMS. http://www.orientechnologies.com/
7. Titan graph database. http://thinkaurelius.github.com/titan/
8. Abiteboul, S., Quass, D., McHugh, J., Widom, J., Wiener, J.: The Lorel query language for semistructured data. Int. J. Digit. Libr. **1**, 68–88 (1997)
9. Angles, R., Gutierrez, C.: Survey of graph database models. ACM Comput. Surv. **40**, 1:1–1:39 (2008)
10. Cesario, N., Pang, A., Singh, L.: Visualizing node attribute uncertainty in graphs. In: SPIE Proceedings on Visualization and Data Analysis (2011)
11. Dimitrov, D., Singh, L., Mann, J.: Comparison queries for uncertain graphs. In: Decker, H., Lhotská, L., Link, S., Basl, J., Tjoa, A.M. (eds.) DEXA 2013, Part II. LNCS, vol. 8056, pp. 124–140. Springer, Heidelberg (2013)
12. Dimitrov, D., Singh, L., Mann, J.: A process-centric data mining and visual analytic tool for exploring complex social networks. In: IDEA (2013)
13. Fortin, S.: The graph isomorphism problem. Technical Report TR96-20, Department of Computer Science, University of Alberta (1996)
14. Güting, R.H.: GraphDB: modeling and querying graphs in databases. In: VLDB (1994)
15. He, H., Singh, A.K.: Graphs-at-a-time: query language and access methods for graph databases. In: ACM SIGMOD (2008)
16. Jin, R., Liu, L., Aggarwal, C.C.: Discovering highly reliable subgraphs in uncertain graphs. In: ACM SIGKDD (2011)
17. Jin, R., Liu, L., Ding, B., Wang, H.: Distance-constraint reachability computation in uncertain graphs. Proc. VLDB Endow. **4**(9), 551–562 (2011)
18. Koch, C.: MayBMS: a system for managing large uncertain and probabilistic databases. In: Aggarwal, C.C. (ed.) Managing and Mining Uncertain Data. Springer, New York (2009)
19. Mann, J., Sargeant, B.L., Watson-Capps, J.J., Gibson, Q.A., Heithaus, M.R., Connor, R.C., Patterson, E.: Why do dolphins carry sponges? PLoS ONE **3**(12), e3868 (2008)
20. Mann, J., Stanton, M., Patterson, E., Bienestock, E., Singh, L.: Social networks reveal cultural behaviour in tool using dolphins. Nature Commun. **3** (2012). http://www.nature.com/ncomms/journal/v3/n7/full/ncomms1983.html
21. Mann, J., Shark Bay Research Team: Shark bay dolphin project (2011). http://www.monkeymiadolphins.org
22. Moustafa, W.E., Kimmig, A., Deshpande, A., Getoor, L.: Subgraph pattern matching over uncertain graphs with identity linkage uncertainty. CoRR, abs/1305.7006 (2013)
23. Papapetrou, O., Ioannou, E., Skoutas, D.: Efficient discovery of frequent subgraph patterns in uncertain graph databases. In: EDBT/ICDT (2011)
24. Potamias, M., Bonchi, F., Gionis, A., Kollios, G.: k-nearest neighbors in uncertain graphs. Proc. VLDB Endow. **3**, 997–1008 (2010)
25. Prud'hommeaux, E., Seaborne, A.: SPARQL query language for RDF. W3C recommendation 15 (2008)
26. Rubner, Y., Tomasi, C., Guibas, L.J.: The earth mover's distance as a metric for image retrieval. Int. J. Comput. Vision **40**, 99–121 (2000)
27. Sen, P., Deshpande, A., Getoor, L.: Prdb: managing and exploiting rich correlations in probabilistic databases. VLDB J. **18**, 1065–1090 (2009). Special issue on uncertain and probabilistic databases
28. Sen, P., Namata, G.M., Bilgic, M., Getoor, L., Gallagher, B., Eliassi-Rad, T.: Collective classification in network data. AI Mag. **29**(3), 93–106 (2008)

29. Sharara, H., Sopan, A., Namata, G., Getoor, L., Singh, L.: G-PARE: a visual analytic tool for comparative analysis of uncertain graphs. In: IEEE VAST (2011)
30. Shasha, D., Wang, J.T.L., Giugno, R.: Algorithmics and applications of tree and graph searching. In: PODS (2002)
31. Singh, L., Beard, M., Getoor, L., Blake, M.B.: Visual mining of multi-modal social networks at different abstraction levels. In: Information Visualization (2007)
32. Singh, S., Mayfield, C., Mittal, S., Prabhakar, S., Hambrusch, S., Shah, R.: Orion 2.0: native support for uncertain data. In: ACM SIGMOD. ACM (2008)
33. Smolker, R.A., Richards, A.F., Connor, R.C., Mann, J., Berggren, P.: Sponge-carrying by Indian Ocean bottlenose dolphins: possible tool-use by a delphinid. Ethology **103**, 454–465 (1997)
34. Wasserman, S., Faust, K.: Social Network Analysis: Methods and Applications. Cambridge University Press, Cambridge (1994)
35. Widom, J.: Trio: a system for data, uncertainty, and lineage. In: Aggarwal, C.C. (ed.) Managing and Mining Uncertain Data. Springer, New York (2009)
36. Yuan, Y., Chen, L., Wang, G.: Efficiently answering probability threshold-based shortest path queries over uncertain graphs. In: Kitagawa, H., Ishikawa, Y., Li, Q., Watanabe, C. (eds.) DASFAA 2010. LNCS, vol. 5981, pp. 155–170. Springer, Heidelberg (2010)
37. Yuan, Y., Wang, G., Chen, L., Wang, H.: Efficient subgraph similarity search on large probabilistic graph databases. PVLDB **5**(9), 800–811 (2012)
38. Yuan, Y., Wang, G., Chen, L., Wang, H.: Efficient keyword search on uncertain graph data. IEEE Trans. Knowl. Data Eng. **25**(12), 2767–2779 (2013)
39. Zhou, H., Shaverdian, A.A., Jagadish, H.V., Michailidis, G.: Querying graphs with uncertain predicates. In: ACM Workshop on Mining and Learning with Graphs (2010)
40. Zhu, Y., Qin, L., Yu, J.X., Cheng, H.: Finding top-k similar graphs in graph databases. In: EDBT (2012)
41. Zou, Z., Gao, H., Li, J.: Discovering frequent subgraphs over uncertain graph databases under probabilistic semantics. In: Proceedings of the 16th ACM SIGKDD International Conference on Knowledge Discovery and Data Mining, KDD 2010, pp. 633–642. ACM, New York (2010)
42. Zou, Z., Li, J., Gao, H., Zhang, S.: Finding top-k maximal cliques in an uncertain graph. In: ICDE (2010)

Fast Disjoint and Overlapping Community Detection

Yi Song[1]([✉]), Stéphane Bressan[1], and Gillian Dobbie[2]

[1] National University of Singapore, Singapore, Singapore
{songyi,steph}@nus.edu.sg
[2] University of Auckland, Auckland, New Zealand
g.dobbie@auckland.ac.nz

Abstract. We propose algorithms for the detection of disjoint and overlapping communities in networks. The algorithms exploit both the degree and clustering coefficient of vertices as these metrics characterize dense connections, which we hypothesize as being indicative of communities. Each vertex independently seeks the community to which it belongs, by visiting its neighboring vertices and choosing its peers on the basis of their degrees and clustering coefficients. The algorithms are intrinsically data parallel. We devise a version for *Graphics Processing Unit* (GPU). We empirically evaluate the performance of our methods. We measure and compare their efficiency and effectiveness to several state-of-the-art community detection algorithms. Effectiveness is quantified by metrics, namely, modularity, conductance, internal density, cut ratio, weighted community clustering and normalized mutual information. Additionally, average community size and community size distribution are measured. Efficiency is measured by the running time. We show that our methods are both effective and efficient. Meanwhile, the opportunity to parallelize our algorithm yields an efficient solution to the community detection problem.

1 Introduction

A community forms when a group of vertices in a network is more interconnected than its vertices are connected to other vertices in the network. The knowledge of such groups or communities helps find efficient ways to distribute and gather information in online social networks for example. Community detection is a useful tool in various fields such as sociology, biology and marketing. In this paper, we propose efficient yet effective algorithms for the detection of communities in networks.

We model a network as a simple graph $G(V, E)$, where V is a set of vertices and E is a set of edges. G is undirected, un-weighted, and has no self-loop. The idea of our method is, for each vertex, to seek the community to which it belongs by visiting its neighbor vertices. Decisions are made based on the degrees, clustering coefficients of the neighbors and the number of common neighbors. Degree

© Springer-Verlag Berlin Heidelberg 2015
A. Hameurlain et al. (Eds.): TLDKS XVIII, LNCS 8980, pp. 153–179, 2015.
DOI: 10.1007/978-3-662-46845-4_6

and clustering coefficient are two importance properties of graph topology. Clustering coefficient measures the cliquishness of neighborhood and thus indicates clustering in the graph locally [21,41].

Our method starts from a micro perspective, which is different from that of previous works such as *GN* [15,31]. Considering the size of networks in modern applications, we try to design a scalable method in order to deal with the large networks within a reasonable time. Therefore, we try to minimize the number of pair-wise computations among vertices. Instead of comparing all pairs of vertices in a graph, we only explore each vertex's immediate neighborhood. Indeed, vertices in the same community are more likely to be neighbors [16]. This significantly reduces the complexity except in the case of dense graphs. In our method, as vertices can independently explore their neighborhood and join a community by following an immediate neighbor, the algorithms are intrinsically data parallel. We devise a parallel algorithm for disjoint community detection and implement it on a *Graphics Processing Unit* (GPU). In the case of overlapping community detection, a vertex is allowed to belong to several communities if strong connections exist between the vertex and any of those communities.

We empirically evaluate the performance of our algorithm with both real world networks and synthetic networks. We evaluate the quality of communities using metrics from different classes [45], as well as one metric recently proposed in [34]. The metrics include modularity, conductance, internal density, cut ratio, weighted community clustering, and *Normalized Mutual Information* [23]. The metrics indicate the community quality from different perspectives. We measure the efficiency by running time. We compare our algorithms with several state-of-the-art algorithms.

This paper is an extension of our prior work [38]. We include an improved algorithm and new experimental results. Particularly, the major extensions are listed as follows.

1. We extend the method [38] to overlapping community detection. It is possible that each vertex belongs to more than one community. The algorithm is proposed in Sect. 3.2.
2. We compare the new algorithm with two state-of-the-art algorithms, a game theory based algorithm and a label propagation based algorithm. The effectiveness and efficiency of the new algorithm are evaluated. We evaluate the quality of communities from various perspectives, including the adaption of the measurements for disjoint communities to the case of overlapping communities. We also design new synthetic data sets with overlapping communities for these experiments. The description for the new synthetic data sets and metrics for experiment are added in Sects. 4.1 and 4.2. The new experimental results for overlapping community detection are shown in Sect. 4.4.

The rest of the paper is organized as follows. Section 2 briefly reviews the related works on community detection methods. Section 3 presents the algorithms we propose. Section 4 shows the experiment setting, experiment results and results analysis. Finally we conclude in Sect. 5. In this paper, we use the term "community" and "cluster" exchangeably.

2   Related Work

Community detection methods can be categorized into several classes.

Several authors [20,22,33,40,47] use random walks. For example, Pons and Latapy [33] use random walk to calculate the similarities, which they call distance between each pair of adjacent vertices, and then use Ward's agglomerative hierarchical clustering approach to find communities. Jin et al. [22] propose an algorithm based on Markov random walk to unfold the communities, and extract them with a cutoff criterion in terms of conductance. Dongen [40] uses Markov Clustering, which simulates the random walks.

Several authors [6,18,19,31,32] focus on *modularity* which is first proposed by Girvan and Newman [15]. Modularity is defined as the number of edges inside groups minus the expected number in an equivalent graph with edges placed at random. An equivalent graph means that the graph has the same number of edges and the same degree distribution. Clauset [6] defines a local measurement of community structure called *local modularity* and proposes an agglomerative algorithm to maximize the *local modularity* of the communities detected. Girvan and Newman [31] propose a divisive method to identify community. The edges with highest betweenness are removed iteratively, thus disconnecting the graph and creating communities. The best partition has the highest *modularity*. Gregory [18] extends Girvan and Newmans' algorithm [15,31] by defining splitting betweenness and allows a vertex to split into multiple copies and to be found in different communities, and thus forms overlapping communities.

Some authors, e.g., in [11,14], use cliques, subgraphs with certain number of vertices and edges between every two vertices. For example, Du et al. [11] use maximal cliques for community detection. The algorithm proposed enumerates all the maximal cliques for finding clustering kernel, assigns the rest vertices to closest kernels, and merges fractional communities. Palla et al. [14] design the clique percolation method which finds all cliques of size k. Communities detected consist of overlapping sets of fully connected subgraphs, union of k-cliques.

The authors of [1,7,36] detect community in an agglomerative way. Ahn et al. [1] define clusters as sets of edges. Their method groups edges with an agglomerative hierarchical clustering technique. Clauset et al. [7] propose a greedy hierarchical agglomerative algorithm. It starts from each vertex being a community and then joins two communities at each iteration. The two communities are selected based on the idea of maximizing modularity increment. They use dendrogram to represent the whole process.

Some methods, such as those presented in [2,3,17,23,25], detect community in a local manner. For example, Baumes et al. [2,3] propose two heuristics to detect locally dense subgraphs as communities. Two subgraphs with significant overlap can be locally optimal and thus are overlapping communities. The first heuristic finds disjoined clusters by deleting high-ranking vertices and then adds the deleted vertices to one or more clusters. The second heuristic starts from randomly chosen seeds and then adds or deletes one vertex at a time until the density metric cannot be further improved. Goldberg et al. [17] propose an additional requirement based on the work in [2,3], which requires the community to be a connected sub-graph, so that the algorithm is able to examine the connectivity of the cluster found. Lancichinetti et al. [23] utilize local expansion and optimization to find communities. Communities are expanded from random seeds until the finiteness function defined reaches locally maximal. This method depends significantly on the design of the fitness function and corresponding parameters. Lancichinetti et al. [25] propose to detect overlapping communities by examining the significance of a cluster with regard to a global null model during the process of community expansion.

Label propagation algorithm for disjoint community detection is extended to overlapping community detection [8,43] by allowing each vertex to have multiple labels instead of only one label. Jierui and Boleslaw [43] propose speaker-listener label propagation algorithm for overlapping community detection. The labels are spread between vertices according to defined pairwise interact rules. After iterations, the probability of having a label for a vertex indicates the membership strength. Michele et al. [8] propose a democratic approach that lets each vertex vote for the communities surrounding it by using label propagation algorithm. Local communities are then merged to global ones.

Besides, Zhang et al. [48] propose a method that combines spectral mapping, fuzzy clustering and the optimization of a quality function. Yan and Gregory [44] propose an optimization for existing community detection algorithms. Pairwise vertex similarities are measured beforehand, and existing algorithms are applied on the graph with the vertex similarities as edge weights. Rosvall and Bergstrom [35] use an information theoretic approach to detect community in weighted and directed network. Nepusz et al. [30] model overlapping community detection as a nonlinear constrained optimization problem that can be solved by simulated annealing methods. Chen et al. [5] propose a game theory based framework. Each vertex is viewed as an agent and is allowed to join and leave communities based on calculated gain and loss, until an equilibrium is reached. Some authors [4,46,49] propose model-based methods which use nonnegative matrix factorization.

Discussion. Our algorithms detect communities locally, but different from the algorithms in the same category, our algorithms is more straightforward. The algorithms discover communities directly based on the intrinsic properties of the graph, i.e. vertex degree, rather than the designed fitness functions.

3   Algorithm

We propose an algorithm that delegates the job of finding communities to individual vertices. Each vertex seeks its community independently. The decisions of which community to join are made based on the degrees and clustering coefficients of neighbors, as well as the number of common immediate neighbors. We hypothesize that vertices tend to join groups with more connections. In other words, the vertices try to attach themselves to dense structures, i.e. structures with more connections among vertices in this structure.

3.1 Fast Disjoint Community Detection

The algorithm starts by calculating the degrees and local clustering coefficient for each vertex (line 1). The local clustering coefficient is defined as

$$cc[i] = \frac{e_{jk} : j, k \in V, e_{jk} \in E}{degree[i] * (degree[i] - 1)}$$

It is the ratio between the number of edges between vertices within its neighborhood and the number of edges that could possibly exist between them. It quantifies how closely the vertex connects with its neighbors.

Algorithm 1. Fast Community Detection

Input: graph $G(V, E)$ with $|V|$ vertices, $|E|$ edges;
Result: Clusters C_i, $i \in (1, 2, ..., k')$

1   Compute degree[v] and cc[v], $v \in V$;
2   **for** *each v* **do**
3   **if** *degree[v]< degree[v_j]* **then**   /* $v_j \in v_{neighbor}$ */
4         $g[v] \leftarrow v_i$, where degree[v_i] = max(degree[v_j]) ;
5   **else**
6         $g[v] = v$;
7   **for** *each v* **do**
8   **if** *g[v] = v and degree[v] = degree[v_i]* **then**
9   **if** *v and v_i has more than half common vertices;*
10   **then**
11            $g[v] \leftarrow v_i$, if v_i has smaller id;
12   **else**
13         $v_g \leftarrow g[v]$;
14         $c1 \leftarrow$ number of common neighbors between v and j;
15         $c2 \leftarrow$ number of common neighbors between v and $(v_{neighbor} \setminus v_g)$;
16   **if** $c1 < c2$ **then**
17            $g[v] \leftarrow v_i$, where degree[v_i] = max(degree[v_j]), $v_j \in (v_{neighbor} \setminus v_g)$
18   **for** *each v* **do**
19   **if** *g[v] $\neq$ v* **then**
20         $i \leftarrow g[v]$;
21   **repeat**
22            $i \leftarrow g[i]$;
23   **until** *g[i] = i* find standalone vertex;
24         $g[v] \leftarrow i$;
25   $k \leftarrow$ different numbers in g[v];
26   **for** *i from 1* **to** k **do**
27   **for** $v \in C_i$ **do**
28   find the cluster C_j where v has the maximum number of immediate neighbors;
29   **if** $i \neq j$ **then**
30   Cluster v into C_j;
31   **Return** $C_i, i \in (1, 2, ..., k')$;

Next, each vertex looks around its immediate neighbors. If the degree of the vertex, for example vertex v, is the largest among its immediate neighbors, vertex v stands alone and does not follow other vertices. If the degree of vertex v is not the largest among its immediate neighbors and itself, vertex v follows the neighbor with the largest degree among v's immediate neighbors (line 2–6). If more than one vertex among the immediate neighbors have the largest degree, then vertex v follows the one with the largest clustering coefficient, compared to other neighbors.

In the second round, each vertex adjusts their decisions (line 7–17). If the standing-alone vertex v has neighbors with the same degree, check the number of common neighbors of vertex v and v's neighbor that has the same degree. If there are enough common neighbors, these two vertices are suggested to be in the same community. If vertex v does not stand alone but follows some neighbor, we check the number of common neighbors vertex v has with the vertex that it follows, and the number of common neighbors it has with the other neighbors. If vertex v has more common neighbors with its other neighbors than the one it follows, then vertex v turns to the vertex with the second largest degree in the neighborhood or stands alone if it itself has the second largest degree.

In the third round, each vertex finalizes the community which it desires to join (line 18–24). If the vertex that vertex v follows is also following vertex v_i, then vertex v also turns to vertex v_i. In the end, each vertex follows a vertex that stands alone. With all the other vertices that follow this vertex, they form a community.

After each vertex chooses its community (line 25), we post-process the memberships to refine the communities (line 26–30). If any vertex has more connections outside the community than inside the community, it changes its membership. This refinement process may change the number of communities from the last step.

The only input of the algorithm is the graph itself. No pre-defined number of communities is needed. In the experiments, the graph is given as an edge list. The output is the communities.

Fig. 1. Example

Figure 1 shows a graph with 8 vertices and 14 edges. After the first round, vertex 2, 3, 4, 5, 6 all follow vertex 1 ($g[1] = 1$, $g[2] = 1$, $g[3] = 1$, $g[4] = 1$, $g[5] = 1$, $g[6] = 1$), while vertex 7 and 8 follow vertex 6 ($g[7] = 6$, $g[8] = 6$). In the second round for each vertex, the status of vertex 1 is unchanged. The status of vertex 2, 3, 4, 5 is also unchanged, because they have more common neighbors with vertex 1, that they follow than with other vertices ({vertex 2, 3, 4, 5}\themselves), vertex 7 and 8 still follow 6, while vertex 6 changes to stand alone instead of following vertex 1 because vertex 6 has more common neighbors with 7 and 8 than with vertex 1. No more changes happen in the third round and the refinement, and thus the final result is that we find two communities: one community is labeled by vertex 1, and has vertex 1, 2, 3, 4, 5; the other community is labeled by vertex 6, and has vertex 6, 7, 8.

We also devise a parallel version. Both the first and second rounds are parallelized. In the first round the vertices look for the vertex with the largest degree in the neighborhood at the same time. In the second round, each vertex makes a decision concurrently. The rest of the algorithm is sequential.

3.2   Fast Overlapping Community Detection

For the case of overlapping communities, we extend *FCD* with modifications in the second round and post-processing, with an additional input parameter θ.

In the second round, each vertex adjusts its decision (line 7–16). If the vertex v does not stand alone but follows some neighbor, and vertex v has more common neighbors with its other neighbors than the one that it follows, then vertex v turns to stand alone so that vertex v leaves the opportunity of finding its communities to the post-processing part. This aims to cluster controversial vertices after other vertices choose their communities, and therefore there are clear local pictures for the controversial vertices to make decisions.

When post-processing the memberships to refine the communities (line 25–29), the number of connections of each vertex v with each cluster is counted. N_i^v is the number of immediate neighbors that v has in C_i, representing the number of connections. For any vertex v, N_{max}^v equals $max(N_i^v)$ where $1 \le i \le k$ (line 27). It is the maximum number of immediate neighbors of vertex v that it has with some cluster. Each vertex is grouped into the cluster with the most connections, and the clusters that have significant number of connections compared with the maximum number, satisfying the criteria of $N_{max}^v - N_i^v \ge \theta$. The parameter θ, overlapping factor, determines the degree of overlapping. If θ equals 0, vertex v is grouped to the clusters that have N_{max}^v connection with v. If θ equals 1, vertex v is grouped to the clusters that have N_{max}^v or $N_{max}^v - 1$ connections with v. The larger θ is, the more clusters one vertex may be clustered into, and thus the more overlapping vertices there are. A vertex changes its membership if the community to which it currently belongs does not have enough connections with it. Note that overlaps may still exist if θ equals 0.

3.3   Complexity Analysis

The time complexity for calculating the clustering coefficient is $\mathcal{O}(n \cdot d^2)$, where n is the number of vertices and d is the average degree of vertices in the graph. The complexity for the first round is $\mathcal{O}(n \cdot d)$. The complexity for the second round is $\mathcal{O}(n \cdot d^2)$. The complexity for the third round is $\mathcal{O}(n^2)$ in the worst case which is very unlikely to happen. The usual complexity for this part is $\mathcal{O}(\alpha \cdot n)$ where α is generally smaller than the graph diameter and presents a value less than 2 in our experiments. The complexity for the refinement is $\mathcal{O}(n \cdot d^2)$. Therefore, the time complexity for the whole algorithm is $\mathcal{O}((d^2 + \alpha) \cdot n)$ in the worst case. For the parallel version, the complexity for the first round is $\mathcal{O}(d)$. The complexity for the second round is $\mathcal{O}(d^2)$. The rest is the same as that of the sequential version. Thus the time complexity for the whole parallel algorithm is $\mathcal{O}(d^2 + \alpha \cdot n)$ in the worst case.

Algorithm 2. Fast Overlapping Community Detection

Input: graph $G(V, E)$, parameter θ;
Result: Clusters C_i, $i \in (1, 2, ..., k')$

1 Compute degree[v] and cc[v], $v \in V$;
2 **for** *each* v **do**
3   **if** *degree[v]< degree[v_j]* **then**   /* $v_j \in v_{neighbor}$ */
4       $g[v] \leftarrow v_i$, where degree[v_i] = max(degree[v_j]) ;
5   **else**
6       $g[v] \leftarrow v$;
7 **for** *each* v **do**
8   **if** *g[v] = v and degree[v] = degree[v_i]* **then**
9   **if** *v and v_i has more than half common vertices;*
10   **then**
11         $g[v] \leftarrow v_i$, if v_i has smaller id;
12   **else**
13      $v_g \leftarrow g[v]$;
14      $c1 \leftarrow$ number of common neighbors between v and j;
15      $c2 \leftarrow$ number of common neighbors between v and $(v_{neighbor} \setminus v_g)$;
16   **if** $c1 < c2$ **then** $g[v] \leftarrow v$;
17 **for** *each* v **do**
18   **if** *g[v] $\neq$ v* **then**
19      $i \leftarrow g[v]$;
20   **repeat**
21         $i \leftarrow g[i]$;
22   **until** *g[i]= i* find standalone vertex;
23      $g[v] \leftarrow i$;
24 $k \leftarrow$ different numbers in g[v];
25 **repeat**
26   **for** *each* v **do**
27   find clusters $\{C_i | N_{max}^v - N_i^v \geq \theta, 1 \leq i \leq k\}$;
28   **if** $v \notin C_i$ **then** Cluster v into C_i
29 **until** *reach equilibrium*;
30 **Return** $C_i, i \in (1, 2, ..., k')$;

The two algorithms can be applied to the networks according to the preliminary knowledge of communities, e.g. whether they are disjoint or overlapped.

4   Experiment

We conduct experiments on both synthetic and real world graphs, including three benchmarks for community detection. We ran the sequential algorithms on an

2.83 GHz Inter Core, 2 Quad CPU machine with 2 GB of main memory under Windows 8 OS. The parallel algorithm ran on the same machine with a GeForce GTX 560 Ti graphics card having 2048 MB of global memory, 8 multiprocessor and 48 CUDA cores per multiprocessor. The algorithms are implemented in Visual C++ 10.0. The parallel algorithm is implemented using the application programming interface CUDA for the C language. CUDA [9], the C language Compute Unified Device Architecture, is provided by NVIDIA and works on NVIDIA graphic cards. The CUDA programming model consists of a sequential host code combined with a parallel kernel code.

We compare our algorithm for disjoint community detection with three state-of-the-art algorithms: *InfoMap* [35], *WalkTrap* [33] and Girvan and Newman (*GN*) [15,31]. *InfoMap* is based on information theory. *Walktrap* is based on random walk. *InfoMap* has been empirically shown to have better performance compared to other algorithms, for community detection [13]. We compare our algorithm for overlapping community detection with two algorithms: *game theory* algorithm and speaker-listener label propagation algorithm (*SLPA*) [43], which show good performance [42,43]. In the experiment, we directly use the original C++ code of the *game-theory* algorithm provided by author of [5] and Java executable file of *SLPA* provided by author of [43].

4.1 Data Sets

We generate a batch of benchmark graphs [24] with known community structure, number of vertices, the average degree, maximum degree, minimum and maximum size of micro and macro community due to the hierarchical structure, and fraction of edges between vertices belonging to the same or different communities (see Table 2). The first set of graphs are generated with 2,000 vertices and different average degrees while the other parameters remain the same. They have no overlapping communities. For overlapping communities, we generate two sets of graphs. The first set of graphs has 10,000 vertices and different average degrees, while the other parameters are the same. Every five graphs have a similar average degree. We run the algorithm on all the graphs and we take and compare the average values. The second set of graphs generated have a varying number of vertices from 10,000 to 50,000, and for every number of vertices, five graphs are generated.

The real-world benchmark graphs used are listed as follows. Among them, Zachary's Karate Club data, American College Football data and Dolphin network are widely used for evaluating community detection algorithms.

Karate Club data is a social network of karate club members studied by the sociologist Wayne Zachary. The network has 34 members (vertices) and they are separated into two different groups due to a controversy between one of the instructors and administrator of the club.

American College Football data is a network with 115 teams (vertices) which are separated into 12 conferences. An edge exists between two vertices if there is a match between two teams. More games happen among teams within the same conference than teams from different conferences.

Dolphin Network is collected by David Lusseaua [28]. The network represents frequent associations between 62 dolphins (vertices) in a community living off Doubtful Sound, New Zealand.

Email-URV data is collected by Guimer et al. [10]. The network contains user-to-user (address- to-address) links from the network of e-mail interchanges among faculty and graduate students at Rovira i Virgili University of Tarragona, Spain. It's available on Alex Arenas website [12].

Arxiv HEP-PH collected by Leskovec et al. [27], is a collaboration network containing scientific collaborations between authors who submitted papers to High Energy Physics. It is available on the SNAP website [37].

Wiki-Vote, collected by Leskove et al. [26], contains user-to-user (who-vote-whom) links from the Wikipedia network. It is available on the SNAP website [37]. Each vertex represents a user. An edge is created from a user to a candidate if a user votes for Wikipedia admin candidates.

Email-Enron data set contains user-to-user (address-to-address) links. It was made public by the Federal Energy Regulatory Commission during its investigations. We obtained it from [37]. Each vertex represents an email address. An edge exists between vertex i and vertex j if address i sends at least one email message to address j.

Epinions data set contains user-to-user (who-trust-whom) links from Epinions network. It was collected by Epinions staff P. Massa. We obtained it from *trustlet* website [29,39]. Each vertex represents a user. An edge corresponds to a trust or distrust statement from one user to another user.

We extract the largest component of the networks that have more than one component. The number of vertices and the number of edges of each data set are listed in Table 1

4.2   Metrics

We use five metrics to qualify the disjoint communities: modularity, conductance, internal density, cut ratio and weighted community clustering. Modularity, conductance, internal density and cut ratio are selected from four classes of metrics for community [45] so that we can eliminate the bias of having only one kind of metric. Weighted community clustering is a recently proposed metric [34].

The **Modularity** [31] is defined as

$$modularity = \frac{1}{2m}\Sigma_{i,j\in V}(A_{ij} - \frac{k_i k_j}{2m})\delta(c_i, c_j)$$

where $A_{ij} = 1$ if i and j are connected, otherwise $A_{ij} = 0$, and $\delta(c_i, c_j) = 1$ if i and j belong to the same cluster, otherwise $\delta(c_i, c_j) = 0$.

The **Conductance** for a set of vertices S is defined as

$$conductance(S) = \frac{c_s}{2m_s + c_s}$$

where $c_s = |(u,v) \in E : u \in S, v \notin S|$. It is the number of edges with one end in the set and the other end outside the set. $m_s = |(u,v) \in E : u \in S, v \in S|$. It is the number of edges in S.

Table 1. Main parameters for generating the benchmark graphs

Maximum degree	60
Minimum for the micro community sizes	10
Maximum for the micro community sizes	100
Number of overlapping vertices	100
Number of memberships of the overlapping vertices	3
Minimum for the macro community size	100
Maximum for the macro community size	200
The fraction of edges between vertices belonging to different macro-communities	0.1
The fraction of edges between vertices belonging to the same macro but not micro community	0.3

The **Internal Density** for a set of vertices S is defined as

$$InternalDensity(S) = \frac{m_s}{n_s(n_s - 1)/2}$$

where m_s is the same as above. n_s is the number of vertices in S. Internal Density is the internal edge density of S.

Table 2. Description of data sets

	Number of vertices	Number of edges
Karate Club	34	78
Dolphin	62	159
American College Football	115	610
Email-URV	1,133	5,451
Wiki-Vote	7,066	100,736
Arxiv HEP-PH	11,204	117,649
Email-Enron	33,696	180,811
Epinions	119,130	704,276

The **Cut Ratio** for a set of vertices S is defined as

$$CutRatio(S) = \frac{c_s}{n_s(n - n_s)}$$

Cut Ratio is the fraction of existing edges out of all possible edges having one end outside the cluster.

The **Weighted Community Clustering** for a community is defined as

$$WCC(S) = \frac{1}{|S|} \sum_{x \in S} f(x, S)$$

where $f(x,S) = \frac{t(x,S)}{t(x,V)} * \frac{vt(x,V)}{|S\backslash x|+vt(x,V\backslash s)}$ if $t(x,V) \neq 0$; $f(x,S) = 0$ if $t(x,V) = 0$. $t(x,S)$ is the number of triangles that vertex x closes with vertices in S and $vt(x,S)$ is the number of vertices of S that form at least one triangle with x.

High modularity suggests dense connections between the vertices within communities but sparse connections between vertices in different communities, while modularity value of zero suggests the connections within communities are no better than those in random graphs which have no community structures. Conductance, internal density and cut ratio measure the quality of communities in term of the internal and external connectivity. WCC measures the community quality based on the close triangles. High WCC suggests higher probability of closed triangles among the vertices within communities than between communities. In our experiments, we take the average of the conductances of communities found for the conductance of the whole network, and it is the same for the other metrics except modularity.

Additionally, we use a widely adopted metric called normalized mutual information (*NMI*) [23] to measure the quality of detected disjoint or overlapping communities and a revised modularity to measure the quality of overlapping communities. The revised **modularity** for overlapping community is defined as:

$$Q_{ov}^E = \frac{1}{2m}\Sigma_c\Sigma_{i,j\in c}[A_{ij} - \frac{k_ik_j}{2m}]\frac{1}{O_iO_j}$$

where O_i is the number of communities to which vertex i belongs, and O_j is the number of communities to which vertex j belongs.

Normalized Mutual Information (*NMI*) of two sets of communities $\{C_1\}$ and $\{C_2\}$ is defined as:

$$NMI(X|Y) = 1 - [H(X|Y) + H(Y|X)]/2,$$

where $H(X)(H(Y))$ is the entropy of the random variable $X(Y)$ associated to the set of community $\{C_1\}(\{C_2\})$, and $H(X|Y)$ is the conditional entropy of X with respect to Y. For a set of overlapping communities $\{C_1\}$, the membership of a vertex v is viewed as a binary array of v_C elements. v_C is the number of the communities, to which vertex v belongs. The kth element of the array is regarded as the realization of a random variable $(X)_k$. The detailed calculation procedure is described in [23]. *NMI* indicates the similarity between two sets communities. It yields values between 0 and 1. Value 1 corresponds to a perfect match. We compute the *NMI* value of the set of communities detected and the known set of communities of the graphs that we generate.

4.3 Experimental Assessment for Disjoint Community Detection

Figure 2 shows the communities found in the Karate Club network by each algorithm. Figure 3 shows the communities found in the Dolphin new network by each algorithm. Vertices of the same color are in the same community.

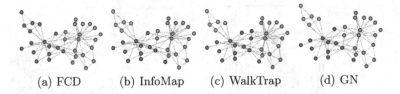

(a) FCD   (b) InfoMap   (c) WalkTrap   (d) GN

Fig. 2. Communities for Karate Club data by different algorithms

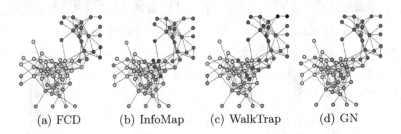

(a) FCD   (b) InfoMap   (c) WalkTrap   (d) GN

Fig. 3. Communities for Dolphin data by different algorithms

Figure 4 shows the measurement results on the four real data sets. The x-axis is labelled by the names of data sets. The y-axis is the value of metric. For each data set, the metric values for the communities detected by each algorithm are compared. Figure 4(a) shows that the communities that FCD and ParallelFCD found have a lower modularity on these four data sets. However, this does not indicate that our algorithm is not better than the other three algorithms. Figure 2 shows that our algorithm identifies two communities, that coincides with the truth that the members of the Karate Club separated into two different groups due to a controversy, and thus the result of our algorithm is actually more reasonable than the other three algorithms even though the modularity values are lower. Figure 4(b) shows the conductance results. The lower the conductance, the better the communities found. In this case, our algorithm has the lowest conductance on two data sets and highest conductance on the other two data sets. Figure 4(c) shows the internal density results. The higher the internal density, the better the communities found. In this case, our algorithm has highest internal density in three of the four data sets, and the lowest in one data set. Figure 4(d) shows the cut ratio results. The lower the cut ratio, the better the communities found. In this case, our algorithm has the lowest cut ratio in one of the four data sets, and the highest in the other three data sets. Figure 4(e) shows the weighted community clustering results. The higher the WCC, the better the communities found [34]. In this case our algorithm has a lower WCC in three of the four data sets. Figure 4(f) shows the running time. For the four data sets, FCD performs the fastest among the algorithms. $ParallelFCD$ performs faster than $InfoMap$, $WalkTrap$ and GN on the Email-URV data. Comparing the performances of the same algorithm on the four data sets, we can see big differences which are due to

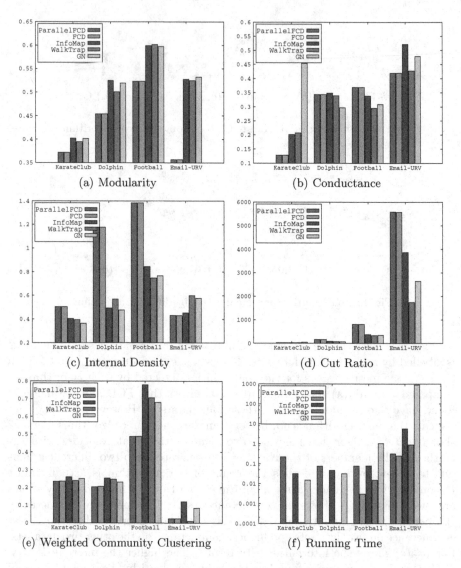

Fig. 4. Measurements on real world graphs

the different graph structures, e.g. different number of vertices, number of edges, different densities.

To sum up the results on these four real data sets, our algorithm, *FCD* and its parallel version, finds communities with better values in terms of internal density and conductance, but not with the other metrics. However, as we can see from the results for Karate Club, the communities detected by our algorithm stay more truthful than those of the other algorithms. In this sense, our algorithm is effective. From the comparison of running time, *FCD* is obviously more efficient than the others.

Figure 5 shows the results on the first set of benchmark graphs. It shows that the metric value changes as the graphs increase in average degree. The x-axis is the average degree of the graphs. The y-axis is the value of metrics. Each dot represents one metric value for the communities detected by one algorithm. Figure 5(a) shows the modularity results. It shows that *WalkTrap* has the highest modularity in general, although in some cases, *GN* and *FCD* have the highest modularity, and *FCD* has a higher modularity than *InfoMap*. Figure 5(b) shows the conductance results. It shows that *InfoMap* has the highest conductance and *GN* has the lowest. Figure 5(c) shows the internal density results. It shows that *InfoMap* has the highest internal density, and *GN* has the lowest density. Figure 5(d) shows the cut ratio results. It shows that *InfoMap* has the highest cut ratio, and *GN* has the lowest. Figure 5(e) shows the *WCC* results. It shows that *FCD* and *WalkTrap* have a higher *WCC*, and *InfoMap* and *GN* have a lower *WCC*. As *FCD* and *ParallelFCD* detect the same communities, the green line and the red line overlap in Fig. 5(a)–(e). Figure 5(f) shows the running time. *FCD* and *ParallelFCD* are shown to be faster in most cases. *GN* is much slower than *InfoMap*, *WalkTrap* and *FCD*. *ParallelFCD* is not obviously faster than *FCD*, due to the data communication between the host *CPU* and device *GPU*. Figure 5(g) shows the measurement of *NMI*. It shows that *InfoMap* and *Walk-Trap* display higher *NMI* values. Figure 5(h) shows the average and deviation of community size. The results reveal that the average size of communities is the closest to the ground truth when the average degree of the graph is about 10 or less than 10. In other words, *FCD* shows better performance in sparse graphs.

Comparing the metric values of the communities found by algorithms and the ground truth, we can see that in some cases *FCD* finds communities closer to the ground truth while in the other cases *GN* and *WalkTrap* find communities closer to the ground truth.

Figure 6 shows the distributions of sizes of communities in four randomly picked graphs. The x-axis is the size of community. The y-axis is the frequency of community size. The results show that *FCD* and *WalkTrap* find communities of closer sizes to the ground truth relatively in general, while in the last case, *GN* finds the communities of the most similar sizes as the known ones.

To sum up the results on these synthetic graphs, *FCD* (*ParallelFCD*) is more stable than *InfoMap* and *GN* in terms of effectiveness. *InfoMap* is the best in terms of internal density but the other three algorithms are better in terms of conductance, cut ratio and *WCC*. *GN* and *WalkTrap* are the best in terms of conductance and cut ratio but the other two algorithms are better in terms of internal density. Comparing the detected communities with the ground truth gives a different evaluation of detected community quality, as the good metric value does not always indicate the closeness of the detected communities to the ground truth. The running time shows that *FCD* is faster than the other three in general.

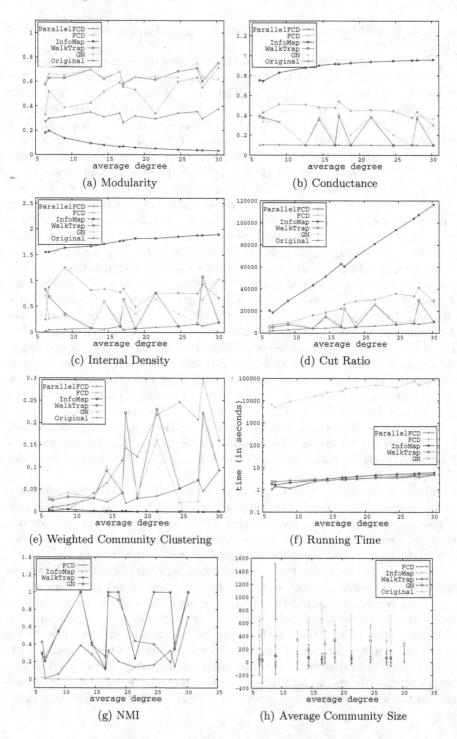

Fig. 5. Measurements on synthetic graphs

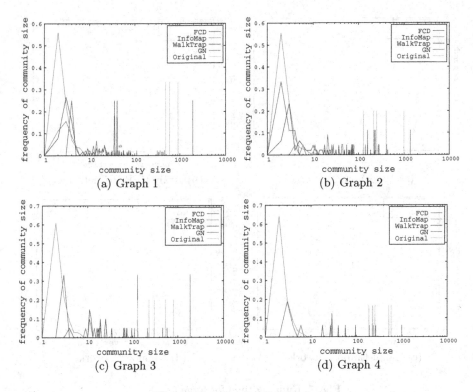

Fig. 6. Community distribution

Another set of experiments demonstrating the running time are carried out on Wiki-Vote, Arxiv HEP-PH, Email-Enron, and Epinion network. We sample subgraphs from the networks. Every subgraph contains k percentage vertices of the original networks, where $k = 10, 20, ..., 90$. We run the *FCD* and *InfoMap* algorithms on these subgraphs and the original graphs. The running time is recorded. Figure 7 shows the running time changing, as the number of vertices of networks increases. Each figure shows the results for one data set. The x-axis is the number of vertices. The y-axis is the time measured in seconds. Due to *WalkTrap* and *GN* algorithms' scalability on large graphs, we only compare the *InfoMap* and *FCD* algorithms here. The results show that both algorithms are able to work with graphs with more than 100,000 vertices. For graphs such as Email-Enron with 33,696 vertices, the algorithms are able to finish the task in a few minutes. In most cases *FCD* is faster than *InfoMap*.

4.4 Experimental Assessment for Overlapping Community Detection

We set the parameter of θ to be 0 in this set of experiments, as we do not expect a large amount of overlaps in our synthetic graphs according to the number of

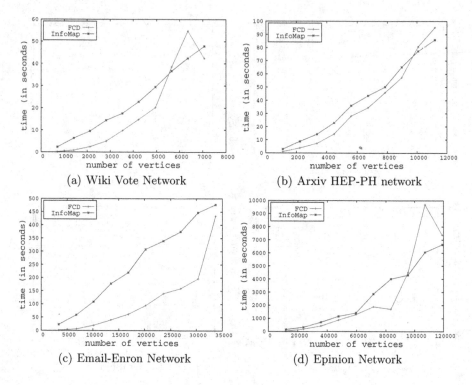

Fig. 7. Running time for large graphs

overlapping vertices and the number of memberships of the overlapping vertices that we set for generating the graphs. We also examined the effects of the higher values of θ and the comparisons indicate a lower quality of detected communities in these graphs when the values of θ is higher.

Figure 8 shows the results for the graphs with varying average degree. The x-axis is the average degree of the graphs. The y-axis is the value of metric. We conduct experiment on five graphs with similar average degree and then take the average of the values to reduce bias against different graph structures. Thus each dot represents one metric value averaged over five values of the communities detected by one algorithm. Figure 8(a) shows the results for *NMI*. It shows that our algorithm *FCD-OV* results in the highest *NMI* value compared to the *GameTheory* and *SLPA* algorithm, which indicates that the communities found by *FCD-OV* are the closest to the true community structure in the input graphs. Figure 8(b)–(f) show the measurement results for community quality. As the community structure of the generated graphs are known, we compare the quality of the communities detected by the three algorithms and the quality of the known communities that is labelled as original in the figures. It is obvious that *FCD-OV* results in the values that are closest to the original ones, suggesting *FCD-OV* has a better capability to find true communities. Figure 8(g) shows the average size of the set of communities found as well as the original

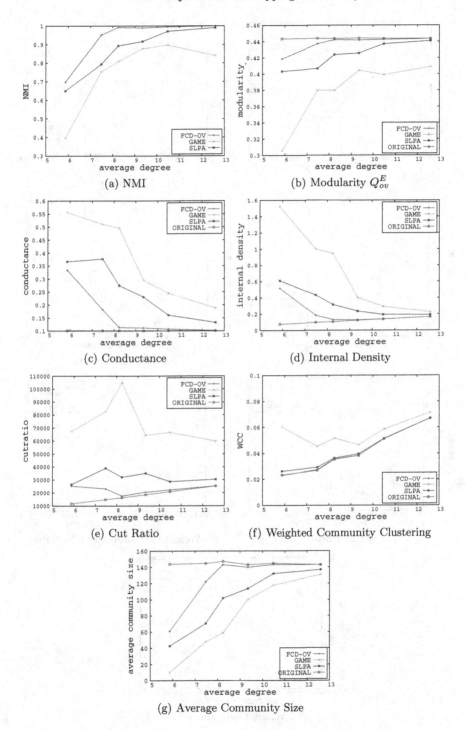

Fig. 8. Measurements on graphs with varying average degree

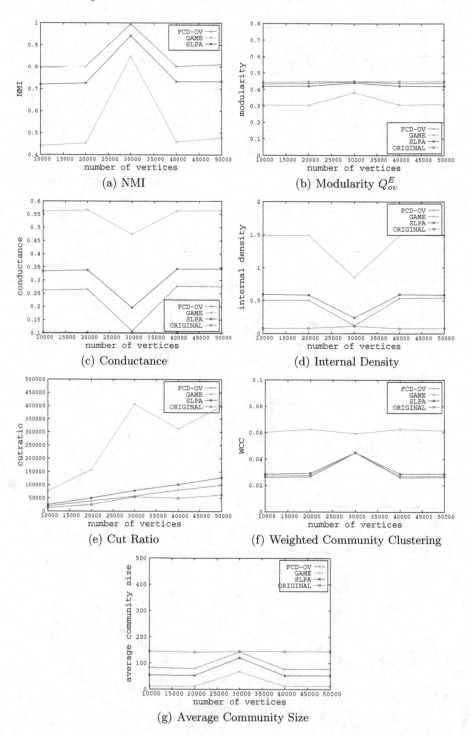

(a) NMI

(b) Modularity Q_{ov}^E

(c) Conductance

(d) Internal Density

(e) Cut Ratio

(f) Weighted Community Clustering

(g) Average Community Size

Fig. 9. Measurements on graphs with varying size

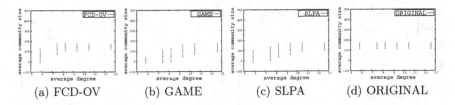

Fig. 10. Measurements on graphs with varying average degree

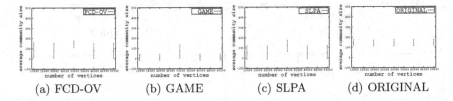

Fig. 11. Measurements on graphs with varying size

average size of the communities in each graph. *FCD-OV* finds the communities with average sizes that are closest to the known ones.

Figure 9 shows the same measurements as Fig. 8 on the graphs with varying size. The x-axis is the number of vertices in the graphs. The y-axis is the value of metric. As the graph size increases from 10,000 to 50,000, we measure the values of each metric for communities found in each graph. We conduct experiment on every five graphs with the same size and then take average of the values to reduce bias against different graph structures. Figure 9(a) shows the results of *NMI*. It shows that *FCD-OV* has the highest values. Figure 9(b)–(f) show the measurement results on community quality. Figure 9(g) shows the average size of the set of communities found as well as the original average size of the communities in each graph.

The results in Fig. 9 suggest that the communities found by *FCD-OV* are the most truthful to the known communities. They also suggest that the change of metric values for the communities is almost independent of the size of graphs except the cut ratio.

Figure 10 shows the average size and standard deviation of each set of the communities in graphs of different average degrees. Figure 11 shows the average size and standard deviation of each set of communities in graphs with different sizes. In Fig. 10, the x-axis is the average degree of the graphs, and the y-axis is the size of community. In Fig. 11, the x-axis is the size of the graph, and the y-axis is the size of the community. It shows that the average size of the communities detected by *FCD-OV* is closer to the average size of known communities.

Figure 12 shows the plots of community distribution for six randomly selected graphs. The x-axis is the size of the community. The y-axis is the ratio of the number of communities of certain size and the total number of communities in the graph, i.e., the frequency of community size. The known communities are

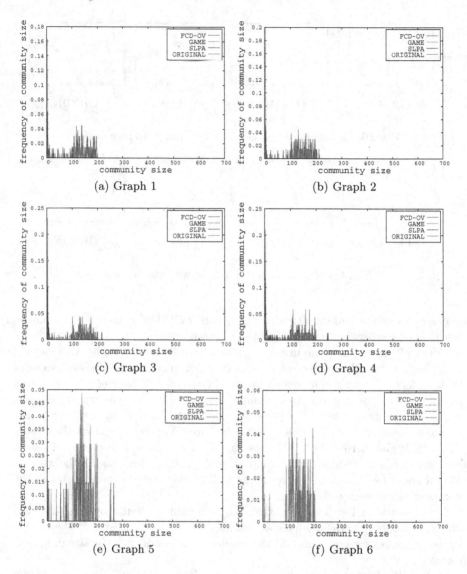

Fig. 12. Community distribution

mostly within size 100 to 200, while many communities detected by the *game-theory* and *SLPA* algorithms are of a size smaller than 100. In comparison, many lines for *FCD-OV* overlap with the lines for the known communities, and this indicates that most of the communities detected by *FCD-OV* are of the sizes of the known communities or close to the sizes of the known communities.

Figure 13 shows the effects of the parameter θ on the values of NMI of the detected communities, and the running time of the *FCD* algorithm. Figure 13(a) shows that, as the value of θ increases, the value of NMI decreases in these cases.

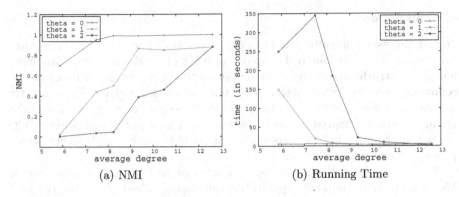

(a) NMI   (b) Running Time

Fig. 13. Effects of θ

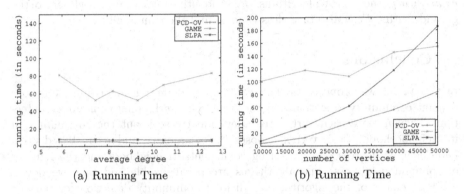

(a) Running Time   (b) Running Time

Fig. 14. Running time comparison

On the other hand, we can see that the effect of θ on the detected communities is larger for graphs with smaller average degrees and smaller for graphs with larger average degrees. Figure 13(b) shows the comparison of running time for different values of θ. The result shows that for graphs with smaller average degrees, a larger input value of θ costs more running time for these graphs. From the results, we can see that an inappropriate value of θ can affect the quality of the resulted communities and increase running time. Therefore, we suggest to choose an appropriate value of θ according to the graphs and the approximated number of overlaps.

Figure 14 shows the running time of three algorithms on the two sets of generated graphs. In both cases, *FCD-OV* costs the least time compared to the *game-theory* and *SLPA* algorithm. The running time also shows that *FCD-OV* detects community in graphs with 50,000 vertices within one and a half minutes. The high efficiency of *FCD-OV* is exhibited.

4.5 Summary

To sum up, we empirically evaluate *FCD* algorithms. For disjoint communities, we examine *FCD* on four real graphs and a set of synthetic graphs. Knowing few ground-truths about the communities in the real graphs, we measure the community quality by calculating the values of chosen metrics. For synthetic graphs, we measure the extend to which the detected communities match the ground-truths. Compared to the *InfoMap*, *WalkTrap* and *GN* algorithms, *FCD* is the fastest and it produces results of comparable quality. *FCD* shows better performance on several metrics.

For overlapping community detection, we examine *FCD* on synthetic graphs. We measure the community quality by calculating values of the metrics and compare the detected communities with the ground-truths. Compared with the *game-theory* and *SLPA* algorithms, *FCD* identifies communities closer to the ground-truths. *FCD* also takes less time to find the communities.

5 Conclusions

In this paper we propose two fast community detection algorithms, one for disjoint community detection and the other for overlapping community detection. They initiate each vertex to independently seek out the community in its neighborhood. Each vertex chooses its community and peers based on a knowledge of degrees and clustering coefficients of neighbors and the number of common neighbors. The algorithms are parallelizable and thus we devise a GPU version of the algorithm for disjoint community detection for parallel computation. In the case of disjoint community detection, we empirically evaluate the performance of *FCD*, and compare it to the *InfoMap*, *WalkTrap* and *GN* algorithms. We find that *FCD* is the fastest, while it produces results of comparable quality. We assess effectiveness based on the values of modularity, conductance, internal density, cut ratio, weighted community clustering, and normalized mutual information as well as community size. In the case of overlapping community detection, we empirically compare the performance of *FCD* for overlapping communities with *game-theory* and *SLPA*. We find that *FCD* for overlapping communities is more efficient, and more effective.

References

1. Ahn, Y.-Y., Bagrow, J.P., Lehmann, S.: Link communities reveal multiscale complexity in networks. Nature **466**, 761 (2010)
2. Baumes, J., Goldberg, M.K., Krishnamoorthy, M.S., Magdon-Ismail, M., Preston, N.: Finding communities by clustering a graph into overlapping subgraphs. In: IADIS AC, pp. 97–104 (2005)
3. Baumes, J., Goldberg, M., Magdon-Ismail, M.: Efficient identification of overlapping communities. In: Kantor, P., Muresan, G., Roberts, F., Zeng, D.D., Wang, F.-Y., Chen, H., Merkle, R.C. (eds.) ISI 2005. LNCS, vol. 3495, pp. 27–36. Springer, Heidelberg (2005)

4. Cao, X., Wang, X., Di, J., Cao, Y., Dongxiao, H.: Identifying overlapping communities as well as hubs and outliers via nonnegative matrix factorization. Scientific report (2013)
5. Chen, W., Liu, Z., Sun, X., Wang, Y.: A game-theoretic framework to identify overlapping communities in social networks. Data Min. Knowl. Discov. **21**(2), 224–240 (2010)
6. Clauset, A.: Finding local community structure in networks. Phys. Rev. E **72**, 026132 (2005)
7. Clauset, A., Newman, M.E.J., Moore, C.: Finding community structure in very large networks. Phys. Rev. E **70**, 066111 (2004)
8. Coscia, M., Rossetti, G., Giannotti, F., Pedreschi, D.: Demon: a local-first discovery method for overlapping communities. CoRR (2012)
9. CUDA-Zone. http://www.nvidia.com/object/what_is_cuda_new.html
10. Danon, L., Diaz-Guilera, A., Giralt, F., Arenas, A.: Self-similar community structure in a network of human interactions. Phys. Rev. E **68**, 065103 (2003)
11. Du, N., Wu, B., Pei, X., Wang, B., Xu, L.: Community detection in large-scale social networks. In: Proceedings of the 9th WebKDD and 1st SNA-KDD 2007 Workshop on Web Mining and Social Network Analysis, WebKDD/SNA-KDD 2007, pp. 16–25. ACM (2007)
12. Email-URV. http://deim.urv.cat/aarenas/data/welcome.htm
13. Fortunato, S., Lancichinetti, A.: Community detection algorithms: a comparative analysis: invited presentation, extended abstract. In: VALUETOOLS 2009. ICST, Brussels, Belgium (2009)
14. Gergely Palla, I.F., Derenyi, I., Vicsek, T.: Uncovering the overlapping community structure of complex networks in nature and society. Nature **435**, 814–818 (2005)
15. Girvan, M., Newman, M.E.J.: Community structure in social and biological networks. Proc. Nat. Acad. Sci. **99**(12), 7821–7826 (2002)
16. Gleich, D.F., Seshadhri, C.: Vertex neighborhoods, low conductance cuts, and good seeds for local community methods. In: Proceedings of the 18th ACM SIGKDD International Conference on Knowledge Discovery and Data Mining, KDD 2012. ACM, New York (2012)
17. Goldberg, M.K., Kelley, S., Magdon-Ismail, M., Mertsalov, K., Wallace, A.: Finding overlapping communities in social networks. In: SocialCom/PASSAT, pp. 104–113 (2010)
18. Gregory, S.: An algorithm to find overlapping community structure in networks. In: Kok, J.N., Koronacki, J., Lopez de Mantaras, R., Matwin, S., Mladenič, D., Skowron, A. (eds.) PKDD 2007. LNCS (LNAI), vol. 4702, pp. 91–102. Springer, Heidelberg (2007)
19. Gregory, S.: A fast algorithm to find overlapping communities in networks. In: Daelemans, W., Goethals, B., Morik, K. (eds.) ECML PKDD 2008, Part I. LNCS (LNAI), vol. 5211, pp. 408–423. Springer, Heidelberg (2008)
20. Harel, D., Koren, Y.: On clustering using random walks. In: Hariharan, R., Mukund, M., Vinay, V. (eds.) FSTTCS 2001. LNCS, vol. 2245, pp. 18–41. Springer, Heidelberg (2001)
21. Holland, P.W., Leinhardt, S.: Transitivity in structural models of small groups. Small Group Res. **2**(2), 107–124 (1971)
22. Jin, D., Yang, B., Baquero, C., Liu, D., He, D., Liu, J.: A markov random walk under constraint for discovering overlapping communities in complex networks. J. Stat. Mech. Theory Exp. **2011**, P05031 (2011)
23. Lancichinetti, A., Fortunato, S., Kertész, J.: Detecting the overlapping and hierarchical community structure in complex networks. New J. Phys. **11**, 033015 (2009)

24. Lancichinetti, A., Fortunato, S., Radicchi, F.: Benchmark graphs for testing community detection algorithms. Phys. Rev. E (Stat. Nonlin. Soft Matter Phys.) **78**(4), 046110 (2008)
25. Lancichinetti, A., Radicchi, F., Ramasco, J.J., Fortunato, S.: Finding statistically significant communities in networks. PLoS One **6**(5)
26. Leskovec, J., Huttenlocher, D., Kleinberg, J.: Predicting positive and negative links in online social networks. In: Proceedings of the 19th International Conference on World Wide Web. ACM (2010)
27. Leskovec, J., Kleinberg, J.M., Faloutsos, C.: Graph evolution: densification and shrinking diameters. TKDD **1**(1), 1–40 (2007)
28. Lusseau, D., Schneider, K., Boisseau, O.J., Haase, P., Slooten, E., Dawson, S.M.: The bottlenose dolphin community of doubtful sound features a large proportion of long-lasting associations. Behav. Ecol. Sociobiol. **54**(4), 396–405 (2003)
29. Massa, P., Avesani, P.: Trust metrics in recommender systems. In: Golbeck, J. (ed.) Computing with Social Trust. Springer, London (2009)
30. Nepusz, T., Petróczi, A., Négyessy, L., Bazsó, F.: Fuzzy communities and the concept of bridgeness in complex networks. Phys. Rev. E **77**(1), 16107 (2008)
31. Newman, M., Girvan, M.: Finding and evaluating community structure in networks. Phys. Rev. E **69**, 026113 (2004)
32. Nicosia, V., Mangioni, G., Carchiolo, V., Malgeri, M.: Extending the definition of modularity to directed graphs with overlapping communities. J. Stat. Mech. Theory Exp. **2009**, P03024 (2009)
33. Pons, P., Latapy, M.: Computing communities in large networks using random walks. In: Yolum, I., Güngör, T., Gürgen, F., Özturan, C. (eds.) ISCIS 2005. LNCS, vol. 3733, pp. 284–293. Springer, Heidelberg (2005)
34. Prat-Pérez, A., Dominguez-Sal, D., Brunat, J.M., Larriba-Pey, J.-L.: Shaping communities out of triangles. In: Proceedings of the 21st ACM International Conference on Information and Knowledge Management, CIKM 2012. ACM (2012)
35. Rosvall, M., Bergstrom, C.T.: Maps of random walks on complex networks reveal community structure. Proc. Nat. Acad. Sci. U.S.A. **105**, 1118–1123 (2008)
36. Schaeffer, S.E.: Graph clustering. Comput. Sci. Rev. **1**(1), 27–64 (2007)
37. SNAP. http://snap.stanford.edu/data
38. Song, Y., Bressan, S.: Fast community detection. DEXA **1**, 404–418 (2013)
39. TrustLet. http://www.trustlet.org/
40. van Dongen, S.M.: Graph clustering by flow simulation. Ph.D. thesis, University of Utrecht (2000)
41. Watts, D.J., Strogatz, S.H.: Collective dynamics of 'small-world' networks. Nature **393**(6684), 409–410 (1998)
42. Xie, J., Kelley, S., Szymanski, B.K.: Overlapping community detection in networks: the state-of-the-art and comparative study. ACM Comput. Surv. **45**(4), 43:1–43:35 (2013)
43. Xie, J., Szymanski, B.K.: Towards linear time overlapping community detection in social networks. In: Tan, P.-N., Chawla, S., Ho, C.K., Bailey, J. (eds.) PAKDD 2012, Part II. LNCS, vol. 7302, pp. 25–36. Springer, Heidelberg (2012)
44. Yan, B., Gregory, S.: Detecting communities in networks by merging cliques. CoRR (2012)
45. Yang, J., Leskovec, J.: Defining and evaluating network communities based on ground-truth. In: Proceedings of the ACM SIGKDD Workshop on Mining Data Semantics, MDS 2012. ACM (2012)
46. Yang, J., Leskovec, J.: Overlapping community detection at scale: a nonnegative matrix factorization approach. In: WSDM (2013)

47. Yen, L., Vanvyve, L., Wouters, D., Fouss, F., Verleysen, F., Saerens, M.: Clustering using a random-walk based distance measure. In: Proceedings of ESANN'2005 (2005)
48. Zhang, S., Wang, R.S., Zhang, X.S.: Identification of overlapping community structure in complex networks using fuzzy c-means clustering. Phys. A **374**(1), 483–490 (2007)
49. Zhang, Z.-Y., Wang, Y., Ahn, Y.-Y.: Overlapping community detection in complex networks using symmetric binary matrix factorization. CoRR (2013)

A Hybrid Approach Using Genetic Programming and Greedy Search for QoS-Aware Web Service Composition

Hui Ma$^{(\boxtimes)}$, Anqi Wang, and Mengjie Zhang

Victoria University of Wellington, Wellington, New Zealand
{hui.ma,mengjie.zhang}@ecs.vuw.ac.nz,
andy.wanganqi@gmail.com

Abstract. Service compositions build new web services by orchestrating sets of existing web services provided in service repositories. Due to the increasing number of available web services, the search space for finding best service compositions is growing exponentially. Further, there are many available web services that provide identical functionality but differ in their Quality of Service (QoS). Decisions need to be made to determine which services are selected to participate in service compositions with optimized QoS properties.

In this paper, a hybrid approach to service composition is proposed that combines the use of genetic programming and random greedy search. The greedy algorithm is utilized to generate valid and locally optimized individuals to populate the initial generation for genetic programming (GP), and to perform mutation operations during genetic programming.

A full experimental evaluation has been carried out using public benchmark test cases with repositories of up to 15,000 web services and 31,000 properties. The results show good performance in searching for best service compositions, where the number of atomic web services used and the tree depth are used as objectives for minimization.

Further, we extend our approach to the more general problem of finding service composition solutions that have near-optimal QoS. Our experimental evaluation demonstrates that our GP-based greedy algorithm enhanced approach can be applied with good performance to the QoS-aware service composition problem.

1 Introduction

Service-oriented software is built on top of service repositories containing hundreds or thousands of atomic web services. In addition to functional properties (i.e., inputs and outputs), web services have non-functional properties, called *quality of service (QoS)*. QoS properties of high practical relevance are for example response time, execution cost, availability and reliability. Even when available web services observe identical or overlapping functionality, they may vary considerably in their QoS properties. To satisfy the functional requirements of a particular service engineering task it is common practice to build suitable composite services by composing atomic web services found in the service repository.

© Springer-Verlag Berlin Heidelberg 2015
A. Hameurlain et al. (Eds.): TLDKS XVIII, LNCS 8980, pp. 180–205, 2015.
DOI: 10.1007/978-3-662-46485-4_7

The *QoS-aware web service composition problem* asks to discover a composite service that satisfies the given functional requirements and has optimal QoS. As more and more web services become available by service providers, the size of service repositories is steadily increasing nowadays. Consequently, the search space for finding the best service composition a particular service engineering task is growing exponentially. Hence, computing an optimal solution is impractical in general. Rather, one is interested in efficient and effective approaches for computing near-optimal solutions.

Web service composition has recently attracted much interest. Many existing approaches tackle service composition tasks by considering them as planning problems using established planning techniques [10,16,22–25]. However, these approaches do not scale. The complexity that they consider is much lower than the one typically observed in service composition tasks based on dedicated web service languages like OWL-S [20] and BPEL4WS [3]. Other approaches tackle service composition tasks by using artificial intelligence techniques [28,30,35,36]. Most approaches have been tested for small service repositories only, without any attention to scalability. In [4,26,29], genetic programming (GP) is used for computing near-optimal service compositions. A thorough analysis reveals the limited effectiveness of the evolutionary process in these GP-based approaches that is due to the complexity of the data structures used and the randomness of the initial population. Therefore, it requires an extremely long time to discover near-optimal solutions, and the results are very unstable, see [26].

Though many works have studied the QoS-aware web service composition problem [2,8,31–34], the process of generating service compositions is generally separated from the process of selecting optimal concrete web services with regard to QoS. The existing approaches assume that the workflow of a service composition is given, and there are many available services that provide identical functions but with different QoS properties. The aim of the existing approach is then to select concrete services for the given workflow so that overall QoS becomes optimal. However, the separation of generating service composition solutions from selecting concrete web services restrict the space of finding optimal service composition solutions. QoS-aware web service composition should consider the generation of a service composition and service selection at the same time. However, finding a service composition with optimal QoS properties is known to be an NP-hard optimization problem [7] and is very time-consuming if the composition process is done manually. GP has shown its ability to provide good approximate solutions to such a problem. Therefore, in this paper we exploit GP to tackle the QoS-aware web service composition problem.

The goal of this paper is to propose a novel GP-based approach to web service composition that overcomes shortcomings of previous GP-based approaches. Our approach to web service composition is a hybrid approach that combines the use of genetic programming and a greedy search algorithm. Instead of starting with an initial population of service compositions that are randomly generated from the huge number of atomic web services in the repository, we apply the greedy search algorithm to pre-filter the repository for those atomic web services that are exclusively related to the given service composition task. We have examined our proposal using the public web service repositories of OWL-S TC [19], WSC2008 [6], and WSC2009 [17] as benchmarks. Further, we extend our work in [27] by adapting our GP-based greedy algorithm enhanced approach to QoS-aware web service composition, and propose two fitness functions to guide the GP-based evolution. We have examined these fitness functions using the service repositories of WSC2008 and WSC2009 extended with QoS properties. We can demonstrate the effectiveness and efficiency of our GP-based greedy algorithm enhanced approach to QoS aware service composition. Specifically, we have investigated the following objectives:

1. Whether the new method can achieve reasonably good performance, and in particular outperforms existing GP-based approaches.
2. Whether the greedy algorithm can effectively discover atomic web services that are exclusively related to the service composition task.
3. Whether the evolved program (the solution to the given service composition task) is interpretable.
4. Whether the new method is suitable for QoS-aware service composition, and which fitness function should be applied during GP evolution.

This paper is structured as follows: Sect. 2 discusses representations of service compositions while Sect. 3 reviews related work on service composition. Sections 4 and 5 present our GP-based approach to the service composition problem. In Sect. 6 we investigate the applicability of our approach to the QoS-aware service composition problem. Section 7 reports on the experiments conducted to test our proposed approaches. Finally, Sect. 8 states our conclusions and suggestions for future research.

2   Representation of Service Compositions

Web services for complex tasks can be composed from atomic web services provided in a service repository. A web service takes certain inputs to generate certain outputs. The inputs and outputs can be semantically described through ontologies as *concepts*, cf. [6,17]. Assume, for example, the given task is to determine the maximum price of a book, its ISBN and the recommended price in dollars for a given input AcademicItemNumber. That is, a web service is needed that takes the concept $I = \{AcademicItemNumber\}$ as input, and produces the concept $O = \{MaxPrice, ISBN, RecommendedPrice\}$ as output, see Fig. 1. If the service repository contains no such atomic web service, a composite service might be able to accomplish the given task.

Service compositions are often represented as directed acyclic graphs, see Fig. 1(a). Squares represent atomic web services used in the composition, while circles represent the input concept I and the output concept O of the composite service S. Arcs are labelled by the properties that are transferred from one atomic service to another, or taken from the input I, or produced for the output O. The service composition must be verified for formal correctness. For that,

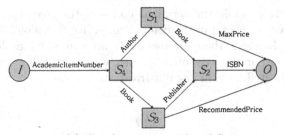

(a) Graph representation of S

	Atomic web service	Input	Output
S_1	GetMaxPrice	Author	Book, MaxPrice
S_2	GetISBN	Book, Publisher	Author, ISBN
S_3	GetRecommendedPricePublisher	Book	RecommendedPrice, Publisher
S_4	GetBookAuthor	AcademicItemNumber	Book, Author

(b) Some atomic web services in the repository

Fig. 1. A composite web service S composed from four atomic web services $S_1, \ldots, S_4$ found in the repository.

each atomic web service used in the composition must satisfy the *matching rule*: its input concept must be subsumed by the union of the properties on its incoming arcs, and its output concept must subsume the union of the properties on its outgoing arcs. In this case, the input concept *matches* the properties received, and the output concept *matches* the properties sent. In our example, all atomic web services satisfy their respective matching rules.

2.1 The Problem

The number of candidate service compositions grows exponentially with the number of atomic web services in the service repository. The problem studied in this paper is how to efficiently find "good" service composition solutions. Due to the inherent complexity of the service composition problem we employ GP to find near-optimal solutions. To do this, we first explore the use of GP for service composition without taking QoS into consideration. Then we extend our approach to QoS-aware service composition. We assume that the reader is familiar with the principles of *genetic programming* (GP) [18]. While directed acyclic graphs are a natural way to represent service compositions, GP traditionally represents evolved programs as tree structures in memory.

2.2 Structures

To facilitate the use of tree-based GP techniques the graph representation of service compositions is converted into a tree representation. We make use of the standard transformation of directed acyclic graphs into trees [5], also known as *unfolding*. In our example, the directed acyclic graph in Fig. 1(a) is transformed into the tree in Fig. 2. Unfolding starts with the output concept O which becomes the root of the tree, while the terminal nodes of the tree represent multiple copies of the input concept I of the composite service S. Unfolding often causes duplicate nodes. In Fig. 2, there are two S_1 nodes, two S_3 nodes, four S_4 nodes, and four I nodes. For the sake of simplicity, we occasionally skip the subtree rooted at a duplicated node in our illustrations.

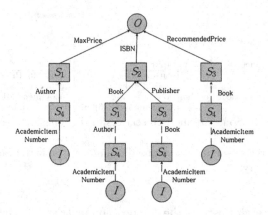

Fig. 2. Tree representation of S

2.3   Genetic Programming Overview

GP [18] is an automated method for creating a working computer program for a problem described by a high-level problem statement. It is considered as a special application of Genetic Algorithm, which is based on Darwinian principles of natural selection. In GP, the term *population* refers to a collection of candidate solutions, called *individuals*, to an optimisation problem. Each of candidate solutions, referred as *chromosome*, is most commonly represented as a tree. The major variations of GP include the terminal set and function set. The terminal set consists of the variables and constants of the program while the function set consists of the functions of the programs.

GP [18] simulates natural evolution and selection of a population to search for an optimal solution. It evolves computer programs, traditionally represented in memory as tree structures. Trees can be easily evaluated in a recursive manner. Every tree node represents an operator function and every terminal node represents an operand. The search starts from a randomly initial population with a defined number of individuals instead of the entire search space. GP uses a *fitness function* to evaluate the degree of how good (or bad) each individual is. It evolves all individuals to generate a new population with a defined number of generations using three genetic operations: reproduction, crossover and mutation. *Reproduction* picks up individuals into a new population without any modification. *Crossover* is applied on an individual by simply switching one of its nodes with another node from another individual in the population. With a tree representation, replacing a node means replacing the entire subtree rooted at this node. *Mutation* affects an individual in the population by replacing a node randomly chosen in the selected individual. GP stops when the best solution is found, or a defined number of individuals in the generation has been reached. To execute a GP-based algorithm we identify the set of terminals, the set of functions, the fitness function and other relevant variables, such as the size of the population and the number of generations, cf. [18].

3   Related Work

Several GP-based approaches for tackling the service composition problem have been presented in the literature. [4] pioneered the use of GP for finding near-optimal web service compositions. They designed genetic operators and fitness functions for assessing composed services, and also compared GP to genetic algorithms (GA) [13]. However, the proposed approach has only been tested on very small repositories. [26] introduced a context-free grammar to randomly initialize the first population and applied a tree structure to represent the chromosome of each individual. The nodes in the tree represent the functions in GP, which are five control structures, while the leaves represent atomic web services. This approach aims to minimize the number of atomic web services used in the service composition. However, due to the structure used to represent service composition, the initial generation often contains many weak individuals, thus makes the whole approach inefficient and leads unstable results. To overcome shortcomings of [26], [29] refined the tree representation, and introduced the service dependency graph for checking the matching rules. A major limitation is that only atomic web services with a single property as output are permitted. In all the above mentioned GP-based approaches, the terminal nodes of the parse tree represent atomic web services and the non-terminal nodes represent various workflow structures. In [26], the authors employ a context-free grammar in order to compose a range of web services with valid structures. However, none of these approaches take into account QoS criteria while discovering service composition solutions. For example, [26] only considers the number of atomic web services used without being concerned about the quality of the composite service.

Other approaches work with graph representations of service compositions [30,35], but are not GP-based. Based on dependency graphs in OWL-S language, [14] used nodes to represent input and output concepts of web services, and edges represent component web services. They employed a Breadth First Search algorithm to search for one path from the input concept to the output concept. If such a path is found, then the execution of the algorithm is stopped and the path is reported as the solution. The problem of this approach is that it does not support complex control flows as in [26]. Meanwhile, even though it can efficiently find a solution for service composition, it does not provide best solutions or near-optimal solutions because the path found first is not necessary the best path of solution. [30] proposes an ant colony algorithm for the web service composition problem. The approach treats each atomic web service as nodes and output-to-input as edges in a graph so that the web service composition problem is transformed into a path searching problem. It searches a path from arrowtails (provided inputs) to arrowheads (desired outputs) through edges (output-to-input) and nodes (atomic services). There are four parameters affecting the ant colony algorithm. However, so far the ant colony algorithm only had some experimental parameters which may not be the best fit for web service compositions. To search the best four parameters used in ant colony algorithm a genetic algorithm is employed. A shortcoming of the approach is that the algorithm is very complicated to use. Also it is evaluated with very small repositories only.

QoS-aware web services composition describes web services in terms of both functional features and non-functional features. References [31] and [32] adopt integer linear programming (ILP) to optimally select component services in which the objective function is defined as a linear composition of multiple QoS constraints. However, the computational time of the ILP based approach will rise exponentially as the number of available web services increases. In order to overcome these problems, [34] puts forward a GA-based approach, which adopts a one-dimensional chromosome-encoded method. However, the shortcoming of this approach is that the length of the chromosome increases as the number of tasks and candidate services increase. [8] proposes a revised encoding method in which each gene of the chromosome represents the task of a composite service, while its value represents a candidate service. However, this encoding method cannot reflect the relationships among component services in a composite service, and the validity of crossover and mutation operations is not checked to assure legal individuals. An improved GA-based approach on the basis of relational matrix encoding method is presented in [33], where the encoding schema is able to express different execution paths. The GA-based approaches assume that a workflow of tasks is given for a service request. However, often the given workflows are not optimized. Also, frequent checking of validity of execution path of individuals after each crossover and mutation operation results in low efficiency of the approach.

This paper proposes an GP-based greedy algorithm enhanced approach with a new representation to tackle the web service composition problem. We first present our approach to web service composition that is able to consider service composition and service selection at the same time, thus overcoming a known weakness of earlier GP-based approaches proposed in the literature. Then we extend our approach to the QoS-aware web service composition problem.

4   The Novel GP-based Approach

As mentioned in Sect. 3, existing GP-based approaches often start with a low quality population at the initial stage. To overcome this matter, we propose a GP-based greedy algorithm enhanced approach that uses a greedy algorithm combined with our GP-based service composition approach. A greedy algorithm can help to search locally optimal solutions, though using a greedy algorithm by itself cannot generate globally optimal solutions for web service composition in general. Rather, we plan to employ a greedy algorithm to generate locally optimal solutions such that the performance of generating a globally optimal solution by GP can be improved. In particular, we propose to use a greedy algorithm to generate individuals that can form the initial population of our GP-based service composition algorithm.

4.1 Variables in Genetic Programming

To apply GP to the service composition problem, the first major step is to define the variables in GP, i.e., to identify the terminal set, the function set, and the fitness function. We will discuss how tree representations of web service compositions will be used for our tree-based GP approach.

Now, we define the variables commonly used in GP, i.e., the terminal set, the function set, and the fitness function [18]. A *service composition task* is defined by an input concept I, an output concept O, and a repository R of atomic web services. We use the atomic web services in the given repository as the *function set* in GP, i.e., we regard the atomic web services as *functions* that map inputs to outputs. GP uses the tree representation discussed above: the internal nodes correspond to functions, all terminal nodes to the input concept I, and the root to the output concept O.

Terminal Set. A service composition task is defined by an input concept I, an output concept O, and a repository of atomic web services. In our approach, all terminal nodes of the tree represent the given input concept I of the service composition task. In Fig. 2, for example, all terminal nodes of the tree represent the input concept I of the desired composite service S.

Function Set. The atomic web services may be regarded as functions that map their inputs to their outputs. We can directly use the atomic web services in the given repository as the function set in GP. In our approach, all nodes of the tree represent functions, except for the terminal nodes that represent the input concept I, and for the root node that represents the output concept O. In Fig. 2, for example, the function set consists of S_1, S_2, S_3, and S_4, which are the atomic web services chosen from the given repository.

Fitness Function. A *fitness function* is used to measure the quality of candidate compositions. How to measure the quality of service composition depends on the task of web service composition. If we do not consider QoS requirements we use the unduplicated number of atomic web services used in a service composition to measure the fitness of a service composition. The fewer atomic web services used in the service composition, the better its performance will be. In addition, we also use the tree depth to measure the fitness of service compositions. The tree depth corresponds to the length of the longest path from the input concept to the output concept. In our example in Fig. 1, the number of features is 4, since $S_1, \ldots, S_4$ are used and duplicates are not considered, and the depth feature is 3. We use the depth feature only to distinguish service compositions with identical number of features. If two service compositions share their number of features, shallow trees are preferred.

If we consider QoS requirements we will use QoS values of composite services to measure the fitness of the composite services. Details of suitable QoS-aware fitness functions will be discussed in Sect. 6.

4.2   Genetic Operators

GP uses the operations *crossover*, *mutation*, and *reproduction* to evolve individuals, i.e., service compositions in our case. To perform *crossover*, we stochastically select two random individuals and check if there is one node representing the same atomic web service in both individuals, and then swap the node together with their subtrees between the two individuals. This guarantees that the matching rules stay satisfied. In Fig. 3, for example, the S_3 nodes in the two individuals are swapped together with the subtrees rooted at them. As usual, the two new individuals generated as offsprings from the two individuals from the previous generation are then included into the next generation.

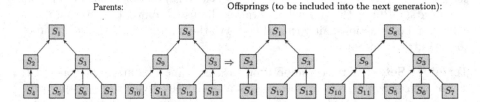

Fig. 3. An example for our crossover.

The *mutation* operator is normally used to replace a node together with its subtree in a selected individual, or to replace only the node. In our approach, we perform mutation by stochastically selecting one node in a randomly chosen individual and replacing its subtree with a new subtree generated by applying a greedy algorithm that will be presented in Sect. 5. In Fig. 4, for example, assume S_3 is selected for mutation and c_a and c_b are properties that S_3 receives from S_5 and S_6, respectively. Then, the mutation operator replaces the subtree of S_3 with a new subtree to generate a new individual as an offspring.

The fitness of the offspring generated by crossover or mutation can be smaller than its parents' one. To avoid a decrease of fitness of the fittest individuals we choose a top percentage of individuals from the old generation for mere *reproduction* and include them into the next generation without any modification.

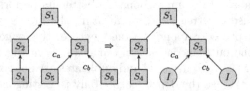

Fig. 4. An example for our mutation.

4.3 GP-based Algorithm for Service Composition

We now present a GP-based algorithm for service composition, see Algorithm 1. The fitness function to be used depends on the requirements of the particular service composition task. In the case of non-QoS aware service composition, the fitness function is presented in Sect. 4.1. For QoS-aware service composition, we will present two fitness functions in Sect. 6.

Input: P /* a set of initial service compositions
Output: an optimal service composition solution S
  Evaluate each individual i in P using the fitness function
  while $g < g_{\max}$ **do**
    Perform reproduction with the rate r
    Select two parents from the population P
    Perform crossover with rate c
    Perform mutation with rate m
    Generate a new population P'
    Evaluate each individual i in P' using fitness function
  end while

Algorithm 1. genetic programming for web services composition

---

As we mentioned in Sect. 2, existing GP-based approaches start with random generated initial populations and therefore take a long time to discover near-optimal solutions. In Sect. 5 we present a greedy algorithm for web service composition to overcome this matter.

5 Random Greedy Search for Initialization and Mutation

Next we propose a randomly greedy algorithm for computing locally optimal solutions for a service composition problem, see Algorithm 2. Its inputs are the input concept I, the output concept O, and the repository R of the service composition task to be solved. The algorithm generates the tree representation of a service composition S that is formally correct for the composition task at hand.

In the algorithm, C_{search} denotes the concept used for searching the repository R, and S_{found} denotes the set of all those atomic web services whose inputs match C_{search}. To begin with, C_{search} is initialized by the input concept I. The discovered atomic web services are added to S_{found}, the outputs of these services are adjoined to C_{search}. Steps 4 to 8 are repeated until no new atomic web service is discovered. In particular this is the case when C_{search} is no longer extended. Afterwards it is checked whether C_{search} subsumes the required output concept O of the composition task. If so, then the composition task has a solution. By applying the matching rule, the nodes of the tree are then stochastically connected to generate the arcs of the tree. Otherwise, there is no solution.

We use the random greedy algorithm as an auxiliary to our GP-based approach in Sect. 4. It generates a set of locally optimal individuals to populate the initial generation for GP. By construction, all of them are formally correct solutions for the composition task at hand. By using the search concept C_{search} the algorithm only considers services that are related to the composition task. The locally optimal individuals constitute an initial population that is already of high quality, thus overcoming a weakness of previous GP-based approaches.

Input: I, O, R
Output: a service composition S, S_{list}
1: $C_{search} \leftarrow I$;
2: $S_{list} \leftarrow \{\}$; /* discover a shrunk set of atomic web services
3: $S_{found} \leftarrow DiscoverService()$;
4: **while** $|S_{found}| > 0$ **do**
5:    $S_{list} \leftarrow S_{list} \cup S_{found}$;
6:    $C_{search} \leftarrow C_{search} \cup C_{output}$ of S_{found};
7:    $S_{found} \leftarrow DiscoverService()$;
8: **end while**
9: **if** $C_{search} \supseteq O$ **then**
10:    $ConnectNodes()$;
11:    Report solution; /* generate a web service composition
12: **else**
13:    Report no solution;
14: **end if**

Algorithm 2. A greedy algorithm for service composition.

Moreover, we apply our greedy algorithm to perform mutation in our GP-based approach. We use it to generate the new subtree rooted at the selected node. This time, the output of the corresponding atomic web service serves as O, and R is restricted to the atomic web services that occur in the composition.

6   QoS-Aware Service Composition

Above we have proposed a novel GP-based greedy algorithm enhanced approach to web service composition. The greedy algorithm is used to reduce the search space of service composition solutions and GP is used to search for service composition solutions that use the smallest number of atomic services. We will now extend our approach to the more general QoS-aware web service composition problem, where additional QoS requirements must be considered.

6.1   QoS Aggregation

The *global QoS* of a composite service is determined by the *local QoS* of its component services and the *composition pattern* of the composite service. To evaluate the global QoS of a composite service, we need to identify a web services quality model so that the global QoS value can be calculated by aggregation from the local QoS values of component services. In accordance with previous works on GP-based web service composition [31] we use the QoS properties availability and reliability, execution cost and response time to exemplify our approach. The values of QoS attributes can either be collected from service providers (e.g. cost), or from records of monitoring previous service execution (e.g. response time, availability). As emphasized in [15,21] these QoS properties represent a selection of relevant characteristics in the field of web services. However, other QoS attributes, e.g. reputation, can be easily included in our approach with only simple modification of the fitness functions that aggregate the values of QoS attributes of composite services.

According to [31], the four parameters are defined as follows. The *availability* A is the probability that a web service is accessible. The global availability of a composite service can be computed as the product of the local availabilities of the atomic web services used in the service composition. For example, the availability of the composite service S in Fig. 5 is $A = a_A \cdot a_B \cdot a_C \cdot a_D \cdot a_E$, where $a_A, \ldots, a_E$ denote the availability of the atomic services $S_A, \ldots, S_E$.

The *reliability* R is the probability that a request is correctly responded within the maximum permitted time frame. The global reliability of a composite service can be computed as the product of the local reliabilities of the atomic web services used in the service composition. For example, the reliability of the composite service S in Fig. 5 is $R = r_A \cdot r_B \cdot r_C \cdot r_D \cdot r_E$, where $r_A, \ldots, r_E$ denote the reliability of the atomic services $S_A, \ldots, S_E$.

The *execution cost* C is the amount of money that a service requester has to pay for executing the web service. The global execution cost of a composite service can be computed as the sum of the execution costs of the atomic web services used in the service composition. For example, the execution of the composite service S in Fig. 5 is $C = c_A + c_B + c_C + c_D + c_E$, where $c_A, \ldots, c_E$ denote the execution cost of the atomic services $S_A, \ldots, S_E$.

The *response time* T is the expected time delay between the moment when a request is sent and the moment when the results are received. When executing a composite web service some of the atomic services can be executed in parallel, while others must be executed in a sequential order. These execution dependencies are reflected by the graph representation of a composite web service.

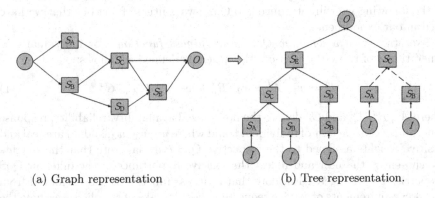

(a) Graph representation   (b) Tree representation.

Fig. 5. A composite web service S composed from five atomic web services $S_A, \ldots, S_E$ found in the repository.

For example, in Fig. 5, the atomic services S_A and S_B can be executed in parallel, as well as S_C and S_D can be executed in parallel. Conversely, S_A must be executed before S_C, and S_C before S_E. Every branch in the tree representation of the composite service S (see Fig. 5) corresponds to a sequence of atomic services that must be executed in the prescribed order. For the aggregation of local QoS values we follow [9] where parallel and sequential execution are considered as the most widely used control structures shared by web service composition languages such as OWL-S and BPEL4WS. The overall time needed for executing the atomic services on a branch b is

$$T_b = \sum_{S_i \text{ is an atomic web service on branch } b} t_i$$

where t_i denotes the response time of S_i. For example, for the left most branch in Fig. 5 the time is $t_A + t_C + t_E$. The global response time of the composite service can then be obtained as the maximum time when ranging over all branches, i.e.,

$$T = \max\{T_b : b \text{ is a branch in the tree representation}\}.$$

Note that while in the tree representation some atomic services are shown multiple times they will only be executed once as illustrated in the graph representation of the composite service.

6.2   QoS-Aware Fitness Functions

To control the GP-based evolution of individuals a suitable fitness function is needed. The fitness of an individual should indicate the goodness of the respective solution for the QoS-aware service composition problem. Hence, we look for a fitness function that reflects the global quality of service of the corresponding service composition, and that can used in our approach as a replacement for the original fitness function introduced in Sect. 4.1. As common practice for the application of GP to multi-criteria optimization problems we will normalize the values of each QoS property considered to be in the interval $[0, 1]$. In the following we will introduce two QoS-aware fitness functions that we have used in our experiments.

We start with a *dynamic QoS-aware fitness function*. For individual i in generation g of the GP evolution, the fitness is defined as follows:

$$fit_i^d = w_1 \cdot A_i^d + w_2 \cdot R_i^d + w_3 \cdot T_i^d + w_4 \cdot C_i^d \tag{1}$$

where A_i^d, R_i^d, T_i^d and C_i^d denote the normalized availability, reliability, response time, and execution cost of individual i, and where w_1, w_2, w_3, and w_4 are real and positive weights assigned to the respective QoS criteria. Note that the weights are chosen by the user and reflect the relative importance of the different QoS properties for the user. Also, note that a fitness function can be achieved from the QoS requirements of service requesters. For the sake of simplicity, assume the weights sum up to 1, that is, $\sum_{j=1}^{4} w_j = 1$.

In GP the fitness of an individual indicates its goodness relative to the other individuals in the same generation. For normalization we therefore use the minimum and maximum values of a particular QoS criteria when ranging over all individuals of the same generation. We use the following formulae to compute the normalized QoS values:

$$A_i^d = \begin{cases} \frac{A_i - A_{min}}{A_{max} - A_{min}} & \text{if } A_{max} - A_{min} \neq 0 \\ 1 & \text{if } A_{max} - A_{min} = 0 \end{cases} \tag{2}$$

$$R_i^d = \begin{cases} \frac{R_i - R_{min}}{R_{max} - R_{min}} & \text{if } R_{max} - R_{min} \neq 0 \\ 1 & \text{if } R_{max} - R_{min} = 0 \end{cases} \tag{3}$$

$$T_i^d = \begin{cases} \frac{T_{max} - T_i}{T_{max} - T_{min}} & \text{if } T_{max} - T_{min} \neq 0 \\ 1 & \text{if } T_{max} - T_{min} = 0 \end{cases} \tag{4}$$

$$C_i^d = \begin{cases} \frac{C_{max} - C_i}{C_{max} - C_{min}} & \text{if } C_{max} - C_{min} \neq 0 \\ 1 & \text{if } C_{max} - C_{min} = 0 \end{cases} \tag{5}$$

Note that the formulae for the availability (A) and reliability (R) slightly differ from the formulae for the response time (T) and execution cost (C). This is because the former two QoS properties are positive criteria, for which the higher the QoS value the better the quality. The latter two QoS properties are negative criteria, for which the lower the QoS value the better the quality. After normalization we have that for each QoS property normalized values closer to 1 indicate better quality, while normalized values closer to 0 indicate worse quality.

For example, to compute the normalized execution cost C_i^d of individual i in some generation we identify the maximum and a minimum value of the execution cost, denoted by C_{max} and C_{min}, across all individuals in the same generation, then we map the minimum to 1 and the maximum to 0. If the original execution cost of individual i is $C_i = 60$ and we find $C_{max} = 100$ and $C_{min} = 50$, then $C_i^d = \frac{C_{max} - C_i}{C_{max} - C_{min}} = 0.8$. So here the execution cost of individual i is quite close to the best execution cost observed among the individuals in the same generation.

Using the dynamic QoS-aware fitness function, QoS-aware web service composition problem is converted into a maximization problem. When the fitness is closer to 1, the solution is more likely to observe better global QoS properties.

We continue with a *static QoS-aware fitness function*. A potential disadvantage of the dynamic fitness function introduced above is that the formulae used for normalization are generation-specific. That is, for each generation we need to find the maximum and minimum values of the QoS properties across all individuals in this generation. This decreases the performance of computing the fitness of individuals. Alternatively, one could use static maximum and minimum values for each QoS property that do not depend on the particular generation in the GP evolution. Indeed, one could determine lower and upper bounds for the global QoS values of composite web services in advance based on the given service repository, and then use them when computing the fitness of individuals during the GP evolution. However, in Sect. 5 we have observed that not all atomic web services in a given repository are related to a given service request. Therefore it is more appropriate to focus on the shrunk repository where all unrelated atomic web services have been deleted (see S_{list} in Algorithm 2).

Let n denote the total number of atomic services in the shrunk repository S_{list}. For the execution cost (C) and response time (T) we determine the smallest values that occur among the atomic web services in the shrunk repository and choose them as C_{min} and T_{min}, respectively. That is, $C_{min} = \min\{c_i : S_i \text{ in } S_{list}\}$ and $T_{min} = \min\{t_i : S_i \text{ in } S_{list}\}$. This gives us lower bounds for the global execution cost and response time of service composition solutions obtained during the GP evolution. For C_{max} and T_{max} we determine the largest values that occur among the atomic web services in the shrunk repository, and multiply them by n. That is, $C_{max} = n \cdot \max\{c_i : S_i \text{ in } S_{list}\}$ and $T_{max} = n \cdot \max\{t_i : S_i \text{ in } S_{list}\}$. This gives us upper bounds for the global execution cost and response time of service composition solutions obtained during the GP evolution.

For the availability (A) and reliability (R) we determine the largest values that occur among the atomic web services in the shrunk repository and choose them as A_{max} and R_{max}, respectively. That is, $A_{max} = \max\{a_i : S_i \text{ in } S_{list}\}$ and $R_{max} = \max\{r_i : S_i \text{ in } S_{list}\}$. This gives us upper bounds for the global availability and reliability of service composition solutions obtained during the GP evolution. For A_{min} and R_{min}, we would in principle determine the smallest values that occur among the atomic web services in the shrunk repository, and compute their nth power. That is, $A_{min} = \min\{a_i : S_i \text{ in } S_{list}\}^n$ and $R_{min} = \min\{r_i : S_i \text{ in } S_{list}\}^n$. This gives us lower bounds for the global availability and reliability of service composition solutions obtained during the GP evolution. For performance reasons we have simplified this to $A_{min} = 0$ and $R_{min} = 0$ in our experiments.

Once we precomputed the lower and upper bounds of global QoS values for composite services for each QoS property under consideration, we can use them to compute the normalized QoS values for the individuals obtained during the GP evolution. For this we use formulae (2) to (5) again, but with the static values A_{min}, A_{max}, R_{min}, R_{max}, T_{min}, T_{max}, C_{min}, C_{max} which do not need to be recomputed for each GP generation. Let A_i^s, R_i^s, T_i^s, C_i^s denote the normalized QoS values obtained for individual i this way. We can then compute the fitness of individual i as follows:

$$fit_i^s = w_1 \cdot A_i^s + w_2 \cdot R_i^s + w_3 \cdot T_i^s + w_3 \cdot C_i^s \tag{6}$$

where w_1, w_2, w_3, and w_4 are real and positive weights with $\sum_{j=1}^4 w_j = 1$ as before.

Recall that in our GP-based approach we always choose a top percentage of individuals from the old generation for mere *reproduction* and include them into the next generation without any modification. Advantage of using the static fitness function is that the fitness of a reproduced individual does not vary from generation to generation. Hence we do not need to recompute the fitness of reproduced individuals, which again increases the performance of the GP evolution.

7   Empirical Results

To evaluate our proposal we tested it using the collection of benchmark test cases provided by the web service competitions WSC2008 and WSC2009. Each test case specifies a service composition task including input concept, output concept, and service repository. The complexity of the composition tasks is very diverse in terms of the overall number of properties considered. In addition, we also used the benchmark test cases of OWL-S TC V2.2 for testing. While the repositories of WSC2008 and WSC2009 are all randomly generated, the repositories of OWL-S TC are all from real domains. The benchmark test suits are publicly available. Some of the test cases of the suites have been used in the literature [26] before, so reusing them allows a comparison. Also, the range of the test cases included represents the diversity of complexity of service composition problems.

Our test platform was a PC with an i5-3320(2.60 GHz) processor, 4.0 GB RAM, and Windows 7 64-bit operating system. As our approach is stochastic, we run each task 30 independent times to record the average and standard deviation of the best fitness and the time consumed. Clearly, the population size and the number of generations used for GP affected the time. For our tests we set the parameters population size = 200, number of generations = 30, reproduction percentage = 0.1, crossover percentage = 0.8, mutation percentage = 0.1, and left the tree depth unbounded.

7.1   Results of the Greedy Algorithm

To evaluate the effectiveness of our greedy algorithm we applied it to all test cases provided by WSC2008 and WSC2009. Table 1 shows the number of atomic web services and the number of properties used for the original repositories, and for the repositories shrunk by applying the greedy algorithm for initialization. For example, for task 1 of WSC2008, there are 158 atomic web services in the original repository, but only 61 of them are related to the given composition task. For task 5 of WSC2009, there are 15211 atomic web services in the original repository, but only 237 of them are related to the composition task. This demonstrates the effectiveness of our greedy algorithm for reducing the search space.

To further evaluate the efficiency of applying our greedy algorithm in service composition we conduct experiments to evaluate the performance of our hybrid approach in comparison to a GP-only approach without greedy search. The experiments are tested on all test cases provided by WSC2008 and WSC2009. For all the test cases both approaches produce the same good results but with different total execution time. The total execution time of the two approaches are shown in the Table 2.

From the experimental results shown above we can see that for all tasks, our hybrid approach uses far less time than the GP-only approach. For some tasks, such as WSC2008-7 and WSC2008-8, our hybrid approach uses less than 10 percent of the time used by the GP-only approach. This justifies the use of random greedy search, and demonstrates its impact on the efficiency of our hybrid approach.

Table 1. Original repositories vs. shrunk repositories from greedy search

Task	Number of atomic web services		Number of properties	
	Original repository	Shrunk repository	Original repository	Shrunk repository
WSC2008-1	158	61	1540	252
WSC2008-2	558	63	1565	245
WSC2008-3	604	106	3089	406
WSC2008-4	1041	45	3135	205
WSC2008-5	1090	103	3067	423
WSC2008-6	2198	205	12468	830
WSC2008-7	4113	165	3075	621
WSC2008-8	8119	132	12337	596
WSC2009-1	572	80	1578	331
WSC2009-2	4129	140	12388	599
WSC2009-3	8138	153	18573	644
WSC2009-4	8301	330	18673	1432
WSC2009-5	15211	237	31044	1025

Table 2. Total time: GP_based composition without Greedy Search vs. GP_based composition with Greedy Search ($\bar{\chi} \pm \sigma$)

Task	Without Greedy Search	With Greedy Search
WSC2008-1	5016 ± 35	2338 ± 33
WSC2008-2	13167 ± 90	2172 ± 23
WSC2008-3	213053 ± 4488	145135 ± 2102
WSC2008-4	10367 ± 104	1301 ± 31
WSC2008-5	48001 ± 222	7630 ± 52
WSC2008-6	328638 ± 8141	38935 ± 297
WSC2008-7	419382 ± 8695	25433 ± 385
WSC2008-8	837021 ± 13187	27123 ± 257
WSC2009-1	828333 ± 78	4886 ± 29
WSC2009-2	397599 ± 4354	19086 ± 93
WSC2009-3	615461 ± 11284	19173 ± 179
WSC2009-4	4198047 ± 42484	192930 ± 3288
WSC2009-5	2953832 ± 158181	93209 ± 382

7.2 Overall Results by GP-Based No-QoS Aware Service Composition

Table 3 shows a comparison of our approach with a recent approach proposed in [26] that also used OWL-S TC, WSC2008, and WSC2009 for testing. Column "Min" records the number of atomic web services in the best known solutions, see [6,17,19]. There are three columns for our approach: Column "Number" records the number of atomic web services in the best solution found by our app-roach, column "Depth" records the tree depth of the best solution, and column "Time" records the search time used for computing the best solution. For the existing approach [26] the respective information is given in the remaining three columns. Note that the search times recorded for the two approaches are not directly comparable as they were evaluated on different platforms. The inten-tion of presenting the time here is to show that our approach is efficient and scalable, as it does not take a long time even for complex tasks using big service repositories.

Our approach was successful in computing a solution for each of the service composition tasks specified by WSC2008 and WSC2009, except for tasks 9 and 10 of WSC2008 which are both known not to have a solution [6]. Recall that our approach only needs the initial greedy search to check for the mere existence of a solution, and is therefore very efficient.

Table 3. Average results for the tests ($\bar{\chi} \pm \sigma$).

Task		Our approach			Existing approach [26]		
Name	Min	Number	Depth	Time (in ms)	Number	Depth	Time (in ms)
OWL-S TC1	1	1.00 ± 0.00	1.00 ± 0.00	14 ± 10	1.00 ± 0.00	1.00 ± 0.00	749 ± 364
OWL-S TC2	2	2.00 ± 0.00	2.00 ± 0.00	51 ± 15	2.00 ± 0.00	2.00 ± 0.00	484 ± 139
OWL-S TC3	2	2.00 ± 0.00	2.00 ± 0.00	250 ± 16	2.00 ± 0.00	2.00 ± 0.00	473 ± 76
OWL-S TC4	4	4.00 ± 0.00	2.63 ± 0.49	341 ± 15	5.70 ± 1.19	2.20 ± 0.40	3010 ± 422
OWL-S TC5	3	3.00 ± 0.00	1.00 ± 0.00	389 ± 21	3.30 ± 0.46	1.00 ± 0.00	1098 ± 240
WSC2008-1	10	10.00 ± 0.00	3.00 ± 0.00	2338 ± 33	15.80 ± 5.71	6.00 ± 1.26	6919 ± 1612
WSC2008-2	5	5.00 ± 0.00	3.87 ± 0.35	2172 ± 23	6.00 ± 0.89	3.50 ± 0.67	11137 ± 3106
WSC2008-3	40	40.60 ± 0.62	23.00 ± 0.00	145135 ± 2102	n/a	n/a	n/a
WSC2008-4	10	10.00 ± 0.00	5.00 ± 0.00	1301 ± 31	n/a	n/a	n/a
WSC2008-5	20	20.00 ± 0.00	8.00 ± 0.00	7630 ± 52	49.90 ± 16.84	9.20 ± 2.96	95390 ± 43521
WSC2008-6	40	45.80 ± 0.92	9.00 ± 0.00	38935 ± 297	n/a	n/a	n/a
WSC2008-7	20	20.00 ± 0.00	15.00 ± 0.00	25433 ± 385	n/a	n/a	n/a
WSC2008-8	30	32.10 ± 0.30	23.00 ± 0.00	27123 ± 257	n/a	n/a	n/a
WSC2008-9	n/a	n/a	n/a	n/a	n/a	n/a	n/a
WSC2008-10	n/a	n/a	n/a	n/a	n/a	n/a	n/a
WSC2009-1	5	5.00 ± 0.00	3.67 ± 0.96	4986 ± 29	n/a	n/a	n/a
WSC2009-2	20	20.03 ± 0.18	6.00 ± 0.00	19086 ± 93	n/a	n/a	n/a
WSC2009-3	10	10.20 ± 0.76	3.07 ± 0.25	19173 ± 179	n/a	n/a	n/a
WSC2009-4	40	42.03 ± 0.85	8.00 ± 4.32	192930 ± 3288	n/a	n/a	n/a
WSC2009-5	30	30.07 ± 0.25	19.00 ± 0.00	93209 ± 382	n/a	n/a	n/a

Note that [26] tested their approach with only 5 tasks from OWL-S TC and 3 tasks from WSC2008. For the first three tasks there is no significant difference between the two approaches. For all the remaining 5 tasks, the statistic significance analysis results shows that our approach is significantly better than the existing approach in [26], i.e., fewer atomic services are used in the best known solutions.

Our approach achieved good test results for the remaining tasks, too. For all solvable tasks, our solutions are interpretable and the numbers of services in our solutions are equal or very close to the numbers of services in the best known solution, with less than 1.00 standard deviation. In terms of search time, our results are also stable with standard deviation less than 5 % of the average. The time consumed is short, even for the most complex tasks of WSC2008 (task 3) and WSC2009 (task 4).

In summary, outcomes of our evaluation show that our hybrid approach to web service composition efficiently generates correct and interpretable near-optimal solutions.

7.3 Examples of Evolved Programs of No-QoS Aware Service Composition

The results of web service composition generated by our GP-based approach are interpretable, i.e., the result trees can be translated into pseudo code that specifies how to compose services step by step. We always start from the leaves (inputs) of the tree and move to the root (output) of the tree. Each node of the tree corresponds to an atomic service, which can be understood as a function call where all its inputs are the function parameters and all its outputs are the function returns.

Due to the page limitation, we only briefly discuss two resulting service composition solutions, one from OWL-S TC V2.2 and one from WSC 2008. This service composition task requests to find a service composition solution for a real world domain problem, i.e., for a given DURATION, CITY and COUNTRY find service composition that provides output WEATHERSEASON, MAP and HOTEL. The service composition solution is very simple, which involves three atomic services processed in parallel, with each of the atomic services producing one of the three outputs required. The tree based result is shown in Fig. 6.

Task OWL-S TC V2.2-5: Get the weather, map and hotel given the city.
Inputs: _CITY, _DURATION, _COUNTRY
Outputs: _WEATHERSEASON, _MAP, _HOTEL
S_1: Service: _CITY_WEATHERSEASON_SERVICE
    Inputs: _CITY
    Outputs: _WEATHERSEASON
S_2: Service: _CITY_MAP_SERVICE
    Inputs: _CITY
    Outputs: _MAP
S_3: Service: _DURATIONCOUNTRYCITY_HOTEL_SERVICE
    Inputs: _CITY, _COUNTRY, _DURATION
    Outputs: _HOTEL

Task 5 of WSC2008 is a relatively complex one. Our greedy algorithm already found a local best solution with 29 atomic web services which, however, is not yet the global best solution. Our GP-based approach then produced the global best solution with 20 atomic web services that are discovered in the first generations. The tree based result is shown in Fig. 7 below.

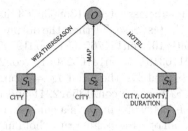

Fig. 6. Service composition solution for task 5 of OWL-S TC V2.2-5.

It can be easily translated into pseudo code. Due to the space limitation we do not show the pseudo code that can be generated from the solution tree. The following shows the task and the solution (shown in Fig. 7) produced by our approach.

Task WSC2008-5: the repository contains 1090 atomic services
Inputs: con428391640, con2100909192
Outputs: con1092196197, con1374634550, con2055848680

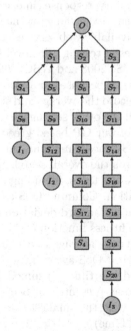

Fig. 7. Service Composition Solution for task 5 of WSC2008-5

The matching rules of atomic web services are validated by checking subsumption. For example, consider task 5 of WSC2008. The input of the task is $I = \{con428391640, con2100909192\}$. It matches the input $I_8 = \{con2100909192, con1368696763\}$ of S_8 because con2100909192 occurs in both I and I_8, while con428391640 in I is a subclass of con1368696763 in I_8 according to the ontology given as part of the test case. Further, the output $O_8 = \{\text{con}1411706461,$ con874272353, con1721591710, con1477657601, con1974742748, con841297848, con1355382428, con2001163191, con1310528051, con2135522241, con78906 9053, con945139087$\}$ of S_8 matches the input $I_4 = \{\text{con}2135522241,\ \text{con}197269$ 4064, con945139087$\}$ of S_4 because con2135522241 and con945139087 occur in both O_8 and I_4, while con789069053 in O_8 is a subclass of con1972694064 in I_4. Similarly one can see that all other matching rules hold in the computed solution.

7.4   Empirical Results of QoS-Aware Service Composition

Above we have seen that our hybrid approach performs well to generate service composition solutions with a small number of atomic web services. In Sect. 6 we have proposed an extension of our approach so that it becomes applicable to the more general QoS-aware service composition problem. To evaluate our proposal we again use the WSC test cases. The service repositories of WSC2008 and WSC2009 have been extended by attaching QoS properties to the description of the atomic web services in the repository. Thus, each atomic web service is characterized by its inputs, outputs and QoS values for the QoS properties considered in this paper (availability, reliability, response time, execution cost). The QoS values have been randomly chosen in the range of the values found in QWS [1], a dataset collected for publicly available web services from real domains. In the following we discuss the outcomes of testing our novel GP-based greedy algorithm enhanced approach with the WSC2008 and WSC2009 repositories extended with QoS properties. Same as in Sect. 7, because our approach is stochastic we run each task 30 independent times to record the average and standard deviation of the best fitness and time consumed. All the other settings are the same as in Sect. 7.

We have conducted tests of our GP-based greedy algorithm enhanced approach using both proposed QoS-fitness functions, the dynamic (fit^d) one and the static one (fit^s). Table 4 records the overall search times used by our algorithm for each of the service composition tasks, including the greedy initialization and the GP evolution till termination. Column "QoS-aware using Dynamic Fitness (Time)" records the average and standard deviation of the time consumed when using the dynamic QoS-aware fitness function (fit^d). Column "QoS-aware using Static Fitness (Time)" records the average and standard deviation of the time consumed when using the static QoS-aware fitness function (fit^s). For comparison we have also included the time consumed by our approach when not considering QoS requirements as presented in Sect. 7 where we just looked for service composition solutions with the smallest number of atomic web services, see column "Not QoS-aware (Time)".

Moreover, we have recorded the average and standard deviation of the QoS value of the best service composition solution found when using the dynamic fitness function (in column "QoS-aware using Dynamic Fitness (Fitness of best)") and when using the static fitness function (in column "QoS-aware using Static Fitness (Fitness of best)") during GP evolution. Note that for better comparison in both cases we show the *static* fitness of the best solutions obtained by the

GP evolution. Recall that the static fitness of an individual is independent of a specific GP generation, but only depends on the shrunk service repository for the particular task. Clearly, for a particular task our approach uses the same shrunk repository, independently of whether the GP evolution is controlled by the dynamic or the static fitness function. Hence, the shown (i.e., static) fitness values of the best solutions for both cases can be directly compared.

7.5 Discussion of Our QoS-Aware Service Composition Approach

In Sect. 6 we have extended our GP-based greedy algorithm enhanced approach to the QoS-aware web service composition problem. A major strength of our approach is that it starts with an initial reduction of the given service repository to those atomic web services that are actually related to the given service composition task. A further strength is the use of a random greedy algorithm to generate the initial population for the GP evolution. Both advantages also hold true for our approach when applied to the more general QoS-aware service composition problem. This initial computation of the shrunk repositories helps to dramatically reduce the search space for the web service composition solutions. Hence, our approach can efficiently find near-optimal service composition solutions in large service repositories with many atomic web services. The greedy algorithm for the random creation of a first population from the shrunk repositories overcomes weaknesses of earlier approaches [26] that suffered from many weak individuals in the beginning of the GP evolution. Further, our approach does not assume a fixed workflow as service composition structure and therefore can search for QoS-optimal service compositions while at the same time selecting QoS-optimal atomic services from the given repository. Furthermore, the use of the greedy algorithm for the initialization and for performing mutations in our approach ensures that each constructed individual constitutes a feasible web service solution. That is, the matching rules derived from the inputs and output

Table 4. Average results of the tests for QoS-aware service composition

Task	Not QoS-aware	QoS-aware using Dynamic Fitness		QoS-aware using Static Fitness	
	Time (in ms)	Time (in ms)	Fitness of best	Time (in ms)	Fitness of best
WSC2008-1	2338 ± 33	2313 ± 73	0.4745 ± 0.0016	2348 ± 62	0.4748 ± 0.0000
WSC2008-2	2172 ± 23	2146 ± 45	0.5168 ± 0.0002	2199 ± 48	0.5415 ± 0.0000
WSC2008-3	145135 ± 2102	182173 ± 2805	0.4253 ± 0.0025	181448 ± 2400	0.4298 ± 0.0013
WSC2008-4	1301 ± 31	1322 ± 88	0.4503 ± 0.0012	1372 ± 75	0.4534 ± 0.0000
WSC2008-5	7630 ± 52	7588 ± 57	0.4656 ± 0.0020	7626 ± 56	0.4683 ± 0.0007
WSC2008-6	38935 ± 297	38697 ± 442	0.4651 ± 0.0018	38376 ± 315	0.4677 ± 0.0010
WSC2008-7	25433 ± 385	25299 ± 394	0.4728 ± 0.0026	25475 ± 364	0.4773 ± 0.0007
WSC2008-8	27123 ± 257	29825 ± 770	0.4512 ± 0.0011	30313 ± 363	0.4532 ± 0.0006
WSC2009-1	4986 ± 29	4889 ± 42	0.5514 ± 0.0000	4974 ± 41	0.5582 ± 0.0110
WSC2009-2	19086 ± 93	19314 ± 164	0.4771 ± 0.0013	19521 ± 175	0.4787 ± 0.0008
WSC2009-3	19173 ± 179	18917 ± 158	0.4908 ± 0.0012	19113 ± 179	0.4911 ± 0.0012
WSC2009-4	192930 ± 3288	190663 ± 2516	0.4775 ± 0.0026	191652 ± 3063	0.4821 ± 0.0006
WSC2009-5	93209 ± 382	89171 ± 1143	0.4687 ± 0.0026	88480 ± 1678	0.4725 ± 0.0008

in the service requests are satisfied by each individual. Therefore, no extra step to check the validity is required and all constructed individuals are interpretable as service composition solutions.

The empirical results of our tests in Sect. 7.4 show that our GP-based greedy algorithm enhanced approach can be extended to the QoS-aware service composition problem without sacrificing much of its performance, when compared to the simpler, not QoS-aware service composition problem. For our tests we have used two different fitness functions to control the GP evolution. Neither of them has emerged as the clear winner over the other. Roughly speaking, for both fitness functions the search time of our approach has the same order of magnitude. The same observation holds true for the QoS of the best individual found in both cases.

For the majority of tasks in Table 4 the search time was slightly less when using the dynamic fitness function and the best individual found had slightly better QoS. However, the observed differences are quite marginal and might be random effects. As argued above, the static fitness function is easier to implement and requires less recomputations during the GP evolution. On the other hand, the dynamic fitness functions permits the use of tighter upper and lower bounds for the QoS values, which are specific for a particular generation in the GP evolution. Thus the dynamic fitness might better capture the goodness of an individual relative to the other individuals in the same generation.

Note that the search time for non-QoS-aware composition, QoS-aware composition with static fitness, and QoS-aware composition with dynamic fitness are not much different. This may be explained as follows: The total search time is determined by the time for initializing individuals, performing crossover and mutation operations, and evaluating the fitness values of individuals. The three kinds of composition just differ in the fitness function that they use. No matter which fitness function is used, the time for evaluating fitness values of individuals are marginal compared to the total search time. In other words, the time used for initializing individuals and for performing crossover and mutation operations dominates the total search time. When comparing the search time for QoS-aware composition with static fitness and that with dynamic fitness we notice the following: while the use of a dynamic fitness function may take extra time for computing the maximum and minimum QoS values within a generation, the individuals in a generation may consist of less atomic services and, therefore, less time is needed overall for calculating the fitness values.

We see no reason to generally favour one QoS-aware fitness function over the other, but recommend to try both for a service composition task at hand. For service composition tasks where good lower and upper bounds for QoS values are unknown or hard to obtain, we recommend the dynamic fitness function, e.g., for service repositories that are not fixed in advance.

8   Conclusions

In this paper we presented an approach for performing web service composition using a combination of GP and greedy search. The random greedy algorithm is an auxiliary to GP. It generates locally optimal individuals for populating the initial generation for GP, and to perform mutations during GP. Moreover, it guarantees that the generated individuals are formally correct and thus interpretable for web service composition. We have applied the GP-based greedy algorithm enhanced approach to service composition without QoS requirements, and also to QoS-aware service composition. We have tested our approach with service composition tasks from the common benchmark test case collections. For QoS-aware service composition we have proposed two QoS-aware fitness functions, a dynamic and a static one. The analysis of the experimental results demonstrates that our approach is efficient, effective and stable for computing near-optimal solutions, when compared to earlier approaches. Most notably, the initial greedy search helps to shrink the number of atomic web services to be considered by GP later on, thus greatly reducing the search space. Our experiments further show that our approach can be extended to QoS-aware service composition using any of the two proposed QoS-aware fitness functions, with a small performance penalty only.

In this paper, multiple QoS criteria are combined into one single criterion to be optimized during the search for service composition solutions. Our service composition approach produces only one close to optimal solution. For the cases where the preferences of QoS properties are not known we may want to provide multiple solutions so that users can choose a solution according to their preferences. Therefore, for future work we will investigate the use of multi-objective GP with the expectation that multiple and often conflicting QoS criteria (e.g., time and cost) can be optimized simultaneously to produce a set of pareto-optimal solutions.

References

1. Al-Masri, E., Mahmoud, Q.H.: QoS-based discovery and ranking of web services. In: IEEE International Conference on Computer Communications and Networks (ICCCN) (2007)
2. Amiri, M.A., Serajzadeh, H.: QoS aware web service composition based on genetic algorithm. In: International Symposium on Telecommunications (IST), pp. 502–507 (2010)
3. Andrews, T.: Business Process Execution Language for Web Services (2003)
4. Aversano, L., di Penta, M., Taneja, K.: A genetic programming approach to support the design of service compositions. Int. J. Comput. Syst. Sci. Eng. 21(4), 247–254 (2006)
5. Bang-Jensen, J.: Digraphs: Theory, Algorithms and Applications. Springer, London (2008)
6. Bansal, A., Blake, M., Kona, S., Bleul, S., Weise, T., Jaeger, M.: WSC-08: Continuing the web services challenge. In: IEEE Conference on E-Commerce Technology, pp. 351–354 (2008)
7. Canfora, G., Di Penta, M.: A lightweight approach for QoS-aware service composition. In: International Conference on Service-Oriented Computing (ICSOC) (2004)

8. Canfora, G., Di Penta, M., Esposito, R., Villani, M.L.: An approach for QoS-aware service composition based on genetic algorithms. In: International Conference on Genetic and Evolutionary Computation (GECCO), pp. 1069–1075 (2005)
9. Cardoso, J., Miller, J., Sheth, A., Arnold, J.: Quality of service for workflows and web service processes. J. Web Semantics 1, 281–308 (2004)
10. Carman, M., Serafini, L., Traverso, P.: Web service composition as planning. In: ICAPS Workshop on Planning for Web Services (2003)
11. Cormen, T.H., Stein, C., Rivest, R.L., Leiserson, C.E.: Introduction to Algorithms. McGraw-Hill, Boston (2001)
12. Elmaghraoui, H., Zaoui, I., Chiadmi, D., Benhlima, L.: Graph-based e-government web service composition. CoRR, abs/1111.6401 (2011)
13. Goldberg, D.E.: Genetic Algorithms in Search, Optimization and Machine Learning. Addison-Wesley, Reading (1989)
14. Hashemian, S., Mavaddat, F.: A graph-based approach to web services composition. In: International Symposium on Applications and the Internet, pp. 183–189 (2005)
15. Jaeger, M.C., Muehl, G.: QoS-based selection of services: The implementation of a genetic algorithm. In: KiVS Workshop on Service-Oriented Architectures and Service-Oriented Computing, pp. 359–370 (2007)
16. Klusch, M., Gerber, A.: Semantic web service composition planning with OWLS-XPlan. In: International AAAI Symposium on Agents and the Semantic Web (2005)
17. Kona, S., et al.: WSC-2009: A quality of service-oriented web services challenge. In: IEEE International Conference on Commerce and Enterprise Computing, pp. 487–490 (2009)
18. Koza, J.: Genetic Programming. MIT Press, Cambridge (1992)
19. Kuster, U., Konig-Ries, B., Krug, A.: An online portal to collect and share SWS descriptions. In: IEEE International Conference on Semantic Computing, pp. 480–481 (2008)
20. Martin, D., et al.: OWL-S Semantic Markup for Web Services (2004)
21. Ma, H., Bastani, F., Yen, I.-L., Mei, H.: QoS-driven service composition with reconfigurable services. IEEE Trans. Serv. Comput. 6(1), 20–34 (2013)
22. Oh, S.-C., Lee, D., Kumara, S.: Effective web service composition in diverse and large-scale service networks. IEEE Trans. Serv. Comput. 1(1), 15–32 (2008)
23. Oh, S.-C., Lee, D., Kumara, S.R.T.: A comparative illustration of AI planning-based web services composition. SIGecom Exch. 5(5), 1–10 (2006)
24. Pistore, M., Marconi, A., Bertoli, P., Traverso, P.: Automated composition of web services by planning at the knowledge level. In: IJCAI, pp. 1252–1259 (2005)
25. Rao, J., Küngas, P., Matskin, M.: Composition of semantic web services using linear logic theorem proving. Inf. Syst. 31(4), 340–360 (2006)
26. Rodriguez-Mier, P., Mucientes, M., Lama, M., Couto, M.: Composition of web services through genetic programming. Evol. Intel. 3, 171–186 (2010)
27. Wang, A., Ma, H., Zhang, M.: Genetic programming with greedy search for web service composition. In: Decker, H., Lhotská, L., Link, S., Basl, J., Tjoa, A.M. (eds.) DEXA 2013, Part II. LNCS, vol. 8056, pp. 9–17. Springer, Heidelberg (2013)
28. Xia, H., Chen, Y., Li, Z., Gao, H., Chen, Y.: Web service selection algorithm based on particle swarm optimization. In: IEEE DASC, pp. 467–472 (2009)
29. Xiao, L., Chang, C., Yang, H.I., Lu, K.S., Jiang, H.Y.: Automated web service composition using genetic programming. In: IEEE COMPSAC, pp. 7–12 (2012)
30. Yang, Z., Shang, C., Liu, Q., Zhao, C.: A dynamic web services composition algorithm. J. Comput. Inf. Syst. 6(8), 2617–2622 (2010)

31. Zeng, L., Benatallah, B., Dumas, M., Kalagnanam, J., Sheng, Q.Z.: Quality driven web services composition. In: International Conference on World Wide Web (WWW), pp. 411–421 (2003)
32. Zeng, L., Benatallah, B., Ngu, A., Dumas, M., Kalagnanam, J., Chang, H.: QoS-aware middleware for web services composition. IEEE Trans. Softw. Eng. **30**(5), 311–327 (2004)
33. Zhang, C., Ma, Y.: Genetic algorithm for QoS-aware web service selection based on chaotic sequences. In: International Conference on Network-Based Information Systems (NBIS), pp. 410–416 (2009)
34. Zhang, L.-J., Li, B.: Requirements driven dynamic services composition for web services and grid solutions. J. Grid Comput. **2**, 121–140 (2004)
35. Zhang, W., Chang, C.K., Feng, T., Jiang, H.Y.: QoS-based dynamic web service composition with ant colony optimization. In: IEEE COMPSAC, pp. 493–502 (2010)
36. Xiangbing, Z., Hongjiang, M., Fang, M.: An optimal approach to the QoS-based WSMO web service composition using genetic algorithm. In: Ghose, A., Zhu, H., Yu, Q., Delis, A., Sheng, Q.Z., Perrin, O., Wang, J., Wang, Y. (eds.) ICSOC 2012. LNCS, vol. 7759, pp. 127–139. Springer, Heidelberg (2013)

Author Index

Printed in the United States
By Bookmasters

Printed in the United States
By Bookmasters